Almost-Periodic Functions and Functional Equations

THE UNIVERSITY SERIES IN HIGHER MATHEMATICS

HALMOS, PAUL R.—Measure Theory
JACOBSON, NATHAN—Lectures in Abstract Algebra
Vol. I—Basic Concepts *Vol. II*—Linear Algebra
Vol. III—Theory of Fields and Galois Theory
KLEENE, S. C.—Introduction to Metamathematics
LOOMIS, LYNN H.—An Introduction to Abstract Harmonic Analysis
LOÈVE, MICHEL—Probability Theory, 3rd Edition
KELLEY, JOHN L.—General Topology
ZARISKI, OSCAR, and SAMUEL, PIERRE—Commutative Algebra, Vols. I and II
GILLMAN, LEONARD, and JERISON, MEYER—Rings of Continuous Functions
RICKART, CHARLES E.—General Theory of Banach Algebras
KELLEY, J. L., NAMIOKA, ISAAC, and CO-AUTHORS—Linear Topological Spaces
SPITZER, FRANK—Principles of Random Walk
SCHENKMAN, EUGENE—Group Theory
NACHBIN, LEOPOLDO—The Haar Integral
KEMENY, JOHN G., SNELL, J. LAURIE, and KNAPP, ANTHONY W.—Denumerable Markov Chains
HÖRMANDER, LARS—An Introduction to Complex Analysis in Several Variables
SARIO, LEO, and NOSHIRO, KIYOSHI—Value Distribution Theory
HALMOS, PAUL R.—A Hilbert Space Problem Book
RODIN, BURTON, and SARIO, LEO—Principal Functions
VARADARAJAN, V. S.—Geometry of Quantum Theory, Vols. I and II
GRÄTZER, GEORGE—Universal Algebra
LUNDELL, ALBERT T., and WEINGRAM, STEPHEN—The Topology of CW Complexes
AMERIO, LUIGI, and PROUSE, GIOVANNI—Almost-Periodic Functions and Functional Equations

A series of advanced text and reference books in pure and applied mathematics. Additional titles will be listed and announced as published.

Almost-Periodic Functions and Functional Equations

LUIGI AMERIO

and

GIOVANNI PROUSE

Istituto di Matematica del Politecnico
Milan, Italy

VNR

Springer Science+Business Media, LLC

ISBN 978-1-4757-1256-8 ISBN 978-1-4757-1254-4 (eBook)
DOI 10.1007/978-1-4757-1254-4

Originally published by Van Nostrand Reinhold Company in 1971

Softcover reprint of the hardcover 1st edition 1971

Library of Congress Catalog Card Number 72–112713

Manufactured in the United States of America.

10 9 8 7 6 5 4 3 2 1

PREFACE

The theory of almost-periodic functions with complex values, created by H. Bohr [1] in his two classical papers published in *Acta Mathematica* in 1925 and 1926, has been developed by many authors and has had noteworthy applications: we recall the works of Weyl, De la Vallée Poussin, Bochner, Stepanov, Wiener, Besicovic, Favard, Delsarte, Maak, Bogoliubov, Levitan. This subject has been widely treated in the monographs by Bohr [2], Favard [1], Besicovic [1], Maak [1], Levitan [1], Cinquini [1], Corduneanu [1], [2]. An important class of almost-periodic functions was studied at the beginning of the century by Bohl and Esclangon.

Bohr's theory has been extended by Muckenhoupt [1] in a particular case and, subsequently, by Bochner [1] and by Bochner and Von Neumann [1] to very general abstract spaces. The extension to Banach spaces is, in particular, of great interest, in view of the fundamental importance of these spaces in theory and application.

Such an extension, which has also been the object of recent research, forms the subject of Part I of this book, in which the almost-periodicity has been considered both in the strong and in the weak sense. Particular importance has been given in this part to the chapter on the integration of almost-periodic functions. This problem, which refers to the very simple equation $x' = f(t)$, serves as a model for the study of much more general almost-periodic equations; in fact, it is apparent already in this case that the statements that can be obtained are very close to those of the theory of numerical almost-periodic functions only for spaces of a rather particular nature. Fortunately, however, such spaces are very important—they include Hilbert spaces and, more generally, uniformly convex spaces.

These considerations are essential in the development of Part II, which concerns the study of some typical equations, linear or nonlinear, of mathematical and theoretical physics. Among the equations we shall consider are the wave equation, Schrödinger's equation with time-dependent operator, and, in the nonlinear field, the wave equation with nonlinear dissipative term and the Navier-Stokes equation.

Although the main object is the study of questions related to almost-periodicity, we have tried to give an adequate account of existence and uniqueness theorems and of the general properties of the solutions.

References relating to concepts of functional analysis have been omitted; for these the reader is to refer, for example, to the books by Day [1], Dunford and Schwartz [1], Hille and Phillips [1], Riesz and Nagy [1], Yoshida [1].

Luigi Amerio
Giovanni Prouse

CONTENTS

PART II
APPLICATIONS TO ALMOST-PERIODIC FUNCTIONAL EQUATIONS

PART I

THEORY OF ALMOST-PERIODIC FUNCTIONS

CHAPTER 1

ALMOST-PERIODIC FUNCTIONS IN BANACH SPACES

1. DEFINITION. ELEMENTARY PROPERTIES

Let X be a Banach space; if $x \in X$, we shall denote by $\|x\|$, or by $\|x\|_X$, the corresponding norm.

Let J be the interval $-\infty < t < +\infty$ and

$$x = f(t) \tag{1.1}$$

a *continuous* function, defined on J and with values in X: a continuous application, in other words, $t \to f(t)$, from J to X.

Continuity will obviously be intended in the *strong sense* (i.e. $\lim_{\tau \to 0} f(t+\tau) = f(t)$ means that $\|f(t+\tau) - f(t)\| \to 0$).

When t varies on J, the point $x = f(t)$ describes, in the X space, a set called the *range* of the function $f(t)$ and denoted by $\mathscr{R}_{f(t)}$.

A set $E \subseteq J$ is said to be *relatively dense* (r.d.) if there exists a number $l > 0$ (*inclusion length*) such that every interval $a \vdash\!\dashv a+l$ contains at least one point of E.

We shall say that the function $f(t)$ is almost-periodic (a.p.) if to every $\varepsilon > 0$ there corresponds an r.d. set $\{\tau\}_\varepsilon$ such that

$$\sup_J \|f(t+\tau) - f(t)\| \leq \varepsilon, \qquad \forall \tau \in \{\tau\}_\varepsilon. \tag{1.2}$$

Each $\tau \in \{\tau\}_\varepsilon$ is called an *ε-almost period* of $f(t)$; to the set $\{\tau\}_\varepsilon$ therefore corresponds an *inclusion length* l_ε and it is clear that, when $\varepsilon \to 0$, the set $\{\tau\}_\varepsilon$ becomes rarefied, whereas, in general, $l_\varepsilon \to +\infty$.

The above definition is a natural extension of Bohr's definition of numerical a.p. function. It is, undoubtedly, in itself a very significant definition: its real depth can actually be understood only *a posteriori* from the beauty of the theory developed from it and the importance of the applications.

The theory of a.p. functions with values in a Banach space, given by Bochner [1], is in its essential lines similar to the theory of numerical a.p. functions; nevertheless new statements arise, as is natural, in connection with questions on compactness and boundedness. These questions, and their developments, are of interest particularly for the integration of a.p. functions and, more generally, for the integration of abstract a.p. equations.

It is obvious that a continuous periodic function is also a.p.

The almost-periodicity condition is, however, much less restrictive than the condition of periodicity; for instance, *all trigonometric polynomials*

$$P(t) = \sum_{1}^{n} {}_k\, a_k e^{i\lambda_k t} \qquad (a_k \in X,\ \lambda_k \in J)$$

are a.p. functions; and, as we shall see later, *the class of a.p. functions coincides with the closure, with respect to the uniform convergence on J, of the set of such polynomials.*

Let us now indicate the first properties of a.p. functions, deducing them directly from their definition. In what follows, when we say that $f(t)$ is uniformly continuous, or bounded, or that the sequence $\{f_n(t)\}$ converges uniformly, etc., we *always* mean that this occurs on the whole interval J.

When, for the sake of clarity, it may be necessary to state in which space $f(t)$ takes its values, we shall say, for instance, that $f(t)$ is X-continuous, or X-a.p., instead of continuous, or a.p., etc.

I. $\forall \varepsilon > 0$, *the set* $\{\tau\}_\varepsilon$ *is closed.*

Let, in fact, τ^* be a limit point for $\{\tau\}_\varepsilon$ and let $\{\tau_n\} \subset \{\tau\}_\varepsilon$ be a sequence such that

$$\lim_{n\to\infty} \tau_n = \tau^*.$$

Then, $\forall t \in J$,

$$\|f(t+\tau_n) - f(t)\| \le \varepsilon$$

and, because of the continuity of $f(t)$,

$$\|f(t+\tau^*) - f(t)\| \le \varepsilon,$$

that is, $\tau^* \in \{\tau\}_\varepsilon$.

II. $\forall \varepsilon > 0$, *the set* $\{l'\}_\varepsilon$ *of the corresponding lengths of inclusion has a minimum*, l_ε.

Let

$$l_\varepsilon = \inf \{l'\}_\varepsilon$$

and consider the interval $a \vdash\!\dashv a + l_\varepsilon$, with arbitrary a. Let τ_n be an ε-a.p. belonging to the interval $a \vdash\!\dashv a + l_\varepsilon + (1/n)$ $(n = 1, 2, \cdots)$ and consider a limit point, τ^*, of the sequence $\{\tau_n\}$; it will then be $a \le \tau^* \le a + l_\varepsilon$. Since we have proved, in theorem I, that τ^* is an ε-a.p., the thesis follows from the fact that a is arbitrary.

Hereafter we may therefore take the value l_ε as length of inclusion corresponding to $\varepsilon > 0$.

III. $f(t)$ *a.p.* $\Rightarrow f(t)$ *uniformly continuous (u.c.).*

It is sufficient to prove that, $\forall \varepsilon > 0$, there exists $\delta_\varepsilon > 0$ such that $\tau \in -\delta_\varepsilon \vdash\!\dashv \delta_\varepsilon \Rightarrow \tau \in \{\tau\}_\varepsilon$. It will then in fact be, $\forall t \in J$ and $\forall \tau \in -\delta_\varepsilon \vdash\!\dashv \delta_\varepsilon$,

$$\|f(t+\tau)-f(t)\| \leq \varepsilon.$$

Let us assume that, for a certain $\sigma > 0$, this does not occur.

There exist then two sequences $\{\delta_n\}$ and $\{t_n\}$, with $\lim_{n\to\infty} \delta_n = 0$, $|\delta_n| \leq 1$, such that

$$\|f(t_n+\delta_n)-f(t_n)\| > \sigma.$$

As $f(t)$ is a.p., there exists, in the interval $-t_n \vdash\!\dashv -t_n + l_{\sigma/4}$, a $\sigma/4$-a.p., τ_n. Consequently,

$$\begin{aligned}
&\|f(t_n+\delta_n+\tau_n)-f(t_n+\tau_n)\| \\
&\quad = \|(f(t_n+\delta_n+\tau_n)-f(t_n+\delta_n))+(f(t_n+\delta_n)-f(t_n))+(f(t_n)-f(t_n+\tau_n))\| \\
&\quad \geq \|f(t_n+\delta_n)-f(t_n)\| - \|f(t_n+\delta_n+\tau_n)-f(t_n+\delta_n)\| \\
&\qquad\qquad - \|f(t_n+\tau_n)-f(t_n)\| > \sigma/2.
\end{aligned}$$

This is absurd; in fact

$$0 \leq t_n+\tau_n \leq l_{\sigma/4}, \quad -1 \leq t_n+\tau_n+\delta_n \leq l_{\sigma/4}+1, \quad \lim_{n\to\infty} \delta_n = 0,$$

and the function $f(t)$ is u.c. in $-1 \vdash\!\dashv l_{\sigma/4}+1$.

IV. $f(t)$ *a.p.* $\Rightarrow \mathscr{R}_{f(t)}$ *relatively compact (r.c.).*

This means that the closure $\overline{\mathscr{R}}_{f(t)}$ is compact. It may be noted that, in the numerical case (or, what is equivalent, if X is Euclidean), statement IV reduces to that of Bohr ($f(t)$ *a.p.* $\Rightarrow f(t)$ *bounded*).

In a general Banach space, however, the r.c. sets are bounded sets of a very particular nature. The assumption that the range $\mathscr{R}_{f(t)}$ is r.c. is equivalent, in fact, to that of the existence, in correspondence to every $\varepsilon > 0$, of a finite number of points:

$$f(t_1), \cdots, f(t_\nu)$$

such that

$$\mathscr{R}_{f(t)} \subset \bigcup_j^{1\ldots\nu} (f(t_j), \varepsilon)$$

(where (x,ε) denotes the open sphere with center x and radius ε); this is also equivalent to the property that every sequence $\{f(s_n)\}$ contains a convergent subsequence (in other words, for the set $\mathscr{R}_{f(t)}$, the Bolzano-Weierstrass principle holds).

To prove the theorem, taken $\varepsilon > 0$ arbitrarily, let l_ε be the corresponding length of inclusion. In $0 \vdash\!\dashv l_\varepsilon$ $f(t)$ is u.c.; there exist therefore ν points

$$f(t_1), \cdots, f(t_\nu) \qquad (0 \leq t_k \leq l_\varepsilon; \quad k = 1, \cdots, \nu)$$

such that

$$f(t) \in \bigcup_j^{1\ldots\nu} (f(t_j), \varepsilon), \qquad \forall t \in 0 \vdash\!\dashv l_\varepsilon.$$

Let t' be an arbitrary point of J. In the interval $-t' \vdash\!\dashv -t'+l_\varepsilon$ there exists an ε-a.p., τ'; therefore

$$\|f(t'+\tau')-f(t')\| \leq \varepsilon, \qquad 0 \leq t'+\tau' \leq l_\varepsilon.$$

Moreover, the point $f(t'+\tau')$ belongs to one of the spheres $(f(t_j), \varepsilon)$; assume $f(t'+\tau') \in (f(t_{j'}), \varepsilon)$. It follows that

$$\|f(t')-f(t_{j'})\| \leq \|f(t')-f(t'+\tau')\| + \|f(t'+\tau')-f(t_{j'})\| < 2\varepsilon,$$

so that

$$f(t') \in \bigcup_{j}^{1\ldots\nu} (f(t_j), 2\varepsilon)$$

and the theorem is proved.

V. $f_n(t)$ *a.p.* $(n=1,2,\cdots)$, $f_n(t) \to f(t)$ *uniformly* $\Rightarrow f(t)$ *a.p.* In other words, *the set of a.p. functions is closed with respect to the topology of uniform convergence.*

Let us fix $\varepsilon > 0$ arbitrarily and choose n_ε in such a way that

$$\sup_J \|f(t)-f_{n_\varepsilon}(t)\| \leq \varepsilon.$$

Let τ be an ε-a.p. for $f_{n_\varepsilon}(t)$. Then, $\forall t \in J$,

$$\begin{aligned}\|f(t+\tau)-f(t)\| &\leq \|f(t+\tau)-f_{n_\varepsilon}(t+\tau)\| + \|f_{n_\varepsilon}(t+\tau)-f_{n_\varepsilon}(t)\| \\ &\quad + \|f_{n_\varepsilon}(t)-f(t)\| \leq 3\varepsilon,\end{aligned}$$

that is, τ is a 3ε-a.p. for $f(t)$.

VI. $f(t)$ *a.p.*, $f'(t)$ *u.c.* $\Rightarrow f'(t)$ *a.p.*

As the derivative $f'(t)$ is u.c. there exists, $\forall \varepsilon > 0$, a $\delta_\varepsilon > 0$ such that, $\forall t''$ and t', with $|t''-t'| \leq \delta_\varepsilon$, we have

$$\|f'(t'')-f'(t')\| \leq \varepsilon.$$

Furthermore, $\forall t \in J$ and for $(1/n) \leq \varepsilon$,

$$\begin{aligned}\left\|n\left(f\left(t+\frac{1}{n}\right)-f(t)\right)-f'(t)\right\| &= \left\|n\int_0^{1/n}(f'(t+\eta)-f'(t))d\eta\right\| \\ &\leq n\int_0^{1/n}\|f'(t+\eta)-f'(t)\|d\eta \leq \varepsilon,\end{aligned}$$

that is, the sequence of a.p. functions $n(f(t+(1/n))-f(t))$ converges uniformly to $f'(t)$. The thesis then follows from theorem V.

VII. $x=f(t)$ *X-a.p.*, $y=g(x)$ *with values in* Y *(Banach space) and continuous on* $\overline{\mathscr{R}}_{f(t)} \Rightarrow g(f(t))$ *Y-a.p.*

Observe, at first, that $g(f(t))$ is continuous. Moreover, $g(x)$ is u.c. on the compact set $\overline{\mathscr{R}}_{f(t)}=G$. Hence, for any $\varepsilon>0$, there exists $\delta_\varepsilon>0$ such that, $\forall x''$ and $x' \in G$, with $\|x''-x'\| \leq \delta_\varepsilon$,

$$\|g(x'')-g(x')\| \leq \varepsilon.$$

Now let τ be a δ_ε-a.p. for $f(t)$. Then, $\forall t$,

$$\|f(t+\tau)-f(t)\| \leq \delta_\varepsilon$$

and consequently (setting $x''=f(t+\tau)$, $x'=f(t)$)

$$\|g(f(t+\tau))-g(f(t))\| \leq \varepsilon.$$

τ is therefore an ε-a.p. for $g(f(t))$ and the thesis is proved.

Corollary. *$f(t)$ a.p., $k>0 \Rightarrow \|f(t)\|^k$ a.p.*

2. BOCHNER'S CRITERION

The a.p. functions have been characterized by Bochner by means of a compactness criterion, which plays an essential role in the theory and in applications. The starting point consists in considering, together with a given function $f(t)$, the set of its translates $\{f(t+s)\}$ and its closure $\overline{\{f(t+s)\}}$ in the topology of uniform convergence.

We shall prove Bochner's criterion by the following analysis.

Let $G=C^0(J;X) \cap L^\infty(J;X)$ be the Banach space of continuous and bounded functions $f(t)$, from J to X, with norm corresponding to uniform convergence; if $\tilde{f}$ is the point of G which corresponds to the function $f(t)$, we have

$$\tilde{f} = \{f(t);\, t\in J\}, \qquad \|\tilde{f}\| = \sup_J \|f(t)\|.$$

Let us now consider, together with $f(t)$, the set of the translates $f(t+s)$, $\forall s \in J$. If

$$\tilde{f}(s) = \{f(t+s);\, t\in J\},$$

we have defined an application, $s \to \tilde{f}(s)$, from J to G; furthermore, $\tilde{f}(0)=\tilde{f}$.

We shall call *transformation of Bochner* the operation by which we pass from $f(t)$ to $\tilde{f}(s)$; $\tilde{f}(s)$ will be called the *Bochner transform* of $f(t)$, using also the notation

$$\tilde{f}(s) = \mathscr{B}(f(t)).$$

The range $\mathscr{R}_{\tilde{f}(s)}$, in G, of the transform $\tilde{f}(s)$, has important properties, expressed by the following statements.

(α) *$\mathscr{R}_{\tilde{f}(s)}$ is a spherical line with center at the origin;* in fact

$$\|\tilde{f}(s)\| = \sup_J \|f(t+s)\| = \sup_J \|f(t)\| = \|\tilde{f}(0)\|. \tag{2.1}$$

(β) *$\mathscr{R}_{\tilde{f}(s)}$ is described in such a way that the "principle of conservation of distance" holds;* in fact

$$\begin{aligned}(2.2)\qquad \|\tilde{f}(s+\tau)-\tilde{f}(s)\| &= \sup_J \|f(t+s+\tau)-f(t+s)\| \\ &= \sup_J \|f(t+\tau)-f(t)\| = \|\tilde{f}(\tau)-\tilde{f}(0)\|.\end{aligned}$$

(γ) *$f(t)$ a.p. $\Leftrightarrow \tilde{f}(s)$ a.p., with the same ε-a.p.* (this is a direct consequence of (2.2)).

(δ) *$\tilde{f}(s)$ a.p. $\Leftrightarrow \mathscr{R}_{\tilde{f}(s)}$ r.c.*

This property, which identifies for the transform $\tilde{f}(s)$ almost-periodicity with relative compactness, is of the greatest importance.

(γ) and (δ) constitute *Bochner's criterion*:

(ε) *$f(t)$ a.p. $\Leftrightarrow \mathscr{R}_{\tilde{f}(s)}$ r.c.*

We shall deduct property (δ) from the following (cf. Amerio [17]):

(ζ) *$\tilde{f}(s)$ a.p. $\Leftrightarrow$ there exist an r.d. sequence $\{s_n\}$ such that the sequence $\{\tilde{f}(s_n)\}$ is r.c.*

By theorem IV the condition is necessary (indeed from that theorem it follows that every sequence $\{\tilde{f}(s_n)\}$ is r.c.).

Let us prove that the condition is sufficient. For this, we shall show first that, $\forall\varepsilon>0$, the set $\{\tau\}_\varepsilon$ of ε-a.p. is r.d. Having assumed that the sequence $\{\tilde{f}(s_n)\}$ is r.c., it is possible to find ν values

$$\tilde{f}(s_{1,0}),\cdots,\tilde{f}(s_{\nu,0})$$

such that, $\forall n$,

$$\tilde{f}(s_n)\in\bigcup_j^{1\ldots\nu}(\tilde{f}(s_{j,0}),\varepsilon).$$

Let us divide the sequence $\tilde{f}(s_n)$ in ν subsequences $\{\tilde{f}(s_{j,n})\}$ such that

$$\tilde{f}(s_{j,n})\in(\tilde{f}(s_{j,0}),\varepsilon),\qquad (j=1,\cdots,\nu).$$

Then

$$\|\tilde{f}(s_{j,n})-\tilde{f}(s_{j,0})\| < \varepsilon$$

that is, by (2.2),

$$\|\tilde{f}(s_{j,n}-s_{j,0})-\tilde{f}(0)\| < \varepsilon.$$

Hence, by (2.2),

$$(2.3)\qquad \tau_{j,n} = s_{j,n}-s_{j,0}$$

is an ε-a.p. for $\tilde{f}(s)$.

We now prove that the sequence $\bigcup_j^{1\ldots\nu}\{\tau_{j,n}\}$ is r.d. Let, in fact, $d>0$ be an inclusion length for the r.d. sequence $\{s_n\}$ and set

$$(2.4)\qquad m = \min_{1\le j\le\nu}\{-s_{j,0}\},\qquad M = \max_{1\le j\le\nu}\{-s_{j,0}\},$$

$$(2.5)\qquad l = M-m+d.$$

Consider an interval $a \vdash\!\dashv a+l$, with arbitrary a. In the interval $a-m \vdash\!\dashv a-m+d$ there falls at least one point, s_{j_1,n_1}, of the sequence $\{s_n\}$; then, by (2.4),

$$a-m+m \leq s_{j_1,n_1}-s_{j_1,0} \leq a-m+d+M,$$

that is, by (2.3),

$$\tau_{j_1,n_1} \in a \vdash\!\dashv a+l,$$

which proves the thesis.

We must now show that $\tilde{f}(s)$ is continuous, that is, by (2.2), that $f(t)$ is u.c.

Setting $\Delta = -d \vdash\!\dashv d$, let $Z=C^0(\Delta; X)$ be the space of functions $z(\eta)$ continuous from Δ to X; if $z \in Z$, we have

$$z = \{z(\eta);\ \eta \in \Delta\}, \qquad \|z\| = \max_{\Delta} \|z(\eta)\|. \tag{2.6}$$

Let us set $z_n=\{f(\eta+s_n);\ \eta \in \Delta\}$ and observe that, as the sequence $\{\tilde{f}(s_n)\}$ is r.c. and

$$\|z_n-z_m\| \leq \|\tilde{f}(s_n)-\tilde{f}(s_m)\|, \tag{2.7}$$

the sequence $\{z_n\}$ is also r.c.

Hence, $f(t)$ being continuous, the functions $f(\eta+s_n)$ are equicontinuous on Δ; $\forall \varepsilon>0$ there exists therefore δ_ε with $0<\delta_\varepsilon \leq (d/2)$ such that, $\forall n$,

$$\eta', \eta'' \in \Delta,\ |\eta'-\eta''| \leq \delta_\varepsilon \Rightarrow \|f(\eta'+s_n)-f(\eta''+s_n)\| \leq \varepsilon. \tag{2.8}$$

For any $\bar{t} \in J$, there exists $s_{\bar{n}}$ in the interval $\bar{t}-(d/2) \vdash\!\dashv \bar{t}+(d/2)$; hence, $\bar{t}=\bar{\eta}+s_{\bar{n}}$, with $|\bar{\eta}| \leq d/2$.

Let $|t-\bar{t}| \leq \delta_\varepsilon$; setting $t=\eta+s_{\bar{n}}$, we obtain $|\eta| \leq d$, $|\eta-\bar{\eta}|=|t-\bar{t}| \leq \delta_\varepsilon$. Then

$$\|f(t)-f(\bar{t})\| = \|f(\eta+s_{\bar{n}})-f(\bar{\eta}+s_{\bar{n}})\| \leq \varepsilon \tag{2.9}$$

and $f(t)$ is therefore u.c.

Observe now that, by (2.3), *it is possible to obtain*, $\forall \varepsilon>0$, *from the given sequence* $\{s_n\}$, *an r.d. sequence of* ε*-a.p.*

If $f(t)$ is a.p., we can take as sequence $\{s_n\}$ that of the integer numbers: $0, \pm 1, \pm 2, \cdots$. It follows that there *exists*, $\forall \varepsilon>0$, *an r.d. sequence of integer* ε*-a.p.*

Note that Bochner's criterion can be stated without introducing the hypothesis that $f(t)$ is *bounded*, as has been done in the definition of the space G. The following statement, in fact, holds.

VIII. *Let $f(t)$ be continuous. $f(t)$ is then a.p. if, and only if, having taken an arbitrary real sequence $\{c_n\}$, there exists a subsequence $\{s_n\}$ such that the sequence $\{f(t+s_n)\}$ converges uniformly.*

That this condition is necessary follows from statement (ε), if we observe that $f(t)$, a.p., is bounded; consequently, the transform $\tilde{f}(s)$ exists.

To show that the condition is sufficient (again basing our proof on (ε)), we observe that, under the assumptions made, $f(t)$ is bounded; in fact, if it were not bounded, there would exist a sequence $\{c_n\}$ such that

$$\lim_{n \to \infty} \|f(c_n)\| = +\infty.$$

No sequence $\{f(t+s_n)\}$, with $\{s_n\} \subseteq \{c_n\}$, could then converge uniformly.

The following important properties follow from Bochner's criterion.

IX. *The sum $f(t)+g(t)$ of two X-a.p. functions is X-a.p.; the product $\varphi(t)f(t)$ of $f(t)$, X-a.p., by a numerical a.p. function $\varphi(t)$ is a.p.*

Hence, the set of X-a.p. functions $f(t)$ is a subspace (closed, by theorem V) of the space G.

Finally, let $X_1, \cdots, X_n$ be Banach spaces, $X = \prod_k^{1 \dots n} X_k$ the (Banach) product space, with $x = \{x_1, \cdots, x_n\}$ and with the norm

$$\|x\| = \sum_{1}^{n}{}_k \|x_k\|$$

(or another equivalent norm).

From Bochner's criterion it follows then that, *if $f_1(t), \cdots, f_n(t)$ are a.p. from J to $X_1, \cdots, X_n$, respectively, the function $x = f(t) = \{f_1(t), \cdots, f_n(t)\}$ is a.p. from J to X.*

Hence, *n a.p. functions admit, $\forall \varepsilon > 0$, a common set $\{\tau\}_\varepsilon$, r.d., of ε-a.p.*

Observation I. Bochner's transform defines a one-to-one correspondence between the functions $f(t) \in G$ and the transforms $\tilde{f}(s) = \mathscr{B}(f(t))$; for these the *principle of the conservation of distance* (property (β)) holds. If we assume that $f(t)$ satisfies a more general condition, but essentially of the same nature as (β), we obtain a sufficient condition for almost-periodicity, given by Bochner [2] in connection with the analysis of the homogeneous wave equation. It is worth noting that this condition is suggested by the *principle of the conservation of energy*, which holds for the solutions of this equation. We shall state this condition in a form corresponding to the statement (ζ).

X. *Assume that $f(t)$, from J to X, satisfies the following conditions:*

(1) *$f(t) \in G$ and there exists an r.d. sequence $\{s_n\}$ such that the sequence $\{f(s_n)\}$ is r.c.*

(2) *$\forall n,m$ the relation*

$$\|f(s_n) - f(s_m)\| \geq \sigma \sup_J \|f(t+s_n) - f(t+s_m)\| \tag{2.10}$$

holds, with $\sigma > 0$, independent of n,m.

Then $f(t)$ is a.p.

From (2.10) it follows in fact $(\forall n,m)$ that

$$\|f(s_n)-f(s_m)\| \geq \sigma\|\tilde{f}(s_n)-\tilde{f}(s_m)\|, \tag{2.11}$$

that is, the sequence $\{\tilde{f}(s_n)\}$ is r.c.

It is clear that (2.10) is verified, with $\sigma=1$, if $f(t)$ satisfies on J the *principle of the conservation of distance*:

$$\|f(s_n)-f(s_m)\| = \|f(t+s_n)-f(t+s_m)\|.$$

OBSERVATION II. Let $f(t)$ be X-a.p. We shall say that *a sequence* $\{s_n\}$ *is regular* (with respect to $f(t)$) *if the sequence* $\{f(t+s_n)\}$ *is uniformly convergent*—in other words, *if the sequence* $\{\tilde{f}(s_n)\}$ *is convergent.*

Let us give a sufficient condition for a sequence $\{s_n\}$ to be regular.

XI. *Let $f(t)$ be a.p. and assume that, for a certain sequence $\{s_n\}$, there exists the limit*

$$\lim_{n\to\infty} f(\eta_k+s_n) = g_k, \tag{2.12}$$

for all the η_k of a sequence $\{\eta_k\}$ dense on J. Then $\{s_n\}$ is regular with respect to $f(t)$.

Suppose, in fact, that the sequence $\{\tilde{f}(s_n)\}$ does not converge. As the range $\mathscr{R}_{\tilde{f}(s)}$ is r.c., there exist two subsequences, $\{s_{n1}\}$ and $\{s_{n2}\}$, of $\{s_n\}$ such that

$$\lim_{n\to\infty} \tilde{f}(s_{n1}) = \tilde{g}_1, \qquad \lim_{n\to\infty} \tilde{f}(s_{n2}) = \tilde{g}_2 \tag{2.13}$$

with

$$\tilde{g}_1 \neq \tilde{g}_2, \tag{2.14}$$

$$\tilde{g}_1 = \{g_1(t);\, t\in J\}, \qquad \tilde{g}_2 = \{g_2(t);\, t\in J\}.$$

As $g_1(t)$ and $g_2(t)$ are continuous and, by (2.12) and (2.13),

$$g_1(\eta_k) = g_2(\eta_k) = g_k,$$

it follows that, $\forall t\in J$,

$$g_1(t) = g_2(t),$$

which contradicts (2.14); hence $\{s_n\}$ is regular.

Let us denote by $\mathscr{S}_f$ the set of *all* sequences regular with respect to $f(t)$. Obviously $\mathscr{S}_f=\mathscr{S}_{\tilde{f}}$.

Given a second function $g(t)$, Y-a.p., *it is of interest* (as we shall see later) *to establish when* $\mathscr{S}_f \subseteq \mathscr{S}_g$.

Let us define first, for the given X-a.p. function $f(t)$, the *translation function* (Bochner [1]):

$$v_f(\tau) = \sup_J \|f(t+\tau)-f(t)\| = \|\tilde{f}(\tau)-\tilde{f}(0)\|. \tag{2.15}$$

Consider, analogously, the function

$$v_g(\tau) = \sup_J \|g(t+\tau)-g(t)\| = \|\tilde{g}(\tau)-\tilde{g}(0)\|.$$

We now define a function $\omega_{f,g}(\varepsilon)$, which we shall call *comparison function* between f and g: $\forall\varepsilon$, with $0<\varepsilon\leq\sup_\tau v_f(\tau)$, we set

$$\omega_{f,g}(\varepsilon) = \sup_{v_f(\tau)\leq\varepsilon} v_g(\tau).$$

Therefore, $\omega_{f,g}(\varepsilon)$ is the supremum of the translation function $v_g(\tau)$ as τ varies on the set of the ε-a.p. of $f(t)$; every ε-a.p. of $f(t)$ is also an $\omega_{f,g}(\varepsilon)$-a.p. of $g(t)$. It follows that $\omega_{f,g}(\varepsilon)$ is a nondecreasing function of ε; hence, there exists the limit

$$\omega_{f,g}(0+) \geq 0. \tag{2.16}$$

The following statement can be proved.

XII. $\mathscr{S}_f \subseteq \mathscr{S}_g \Leftrightarrow \omega_{f,g}(0+)=0.$

Therefore, *a necessary and sufficient condition for the sequences that are regular with respect to $f(t)$ to be also regular with respect to $g(t)$ is that the comparison function $\omega_{f,g}(\varepsilon)$ be infinitesimal with ε*; in other words, it is required that, *given any sequence $\{\tau_n\}$ of ε_n-a.p. for $f(t)$, with $\varepsilon_n\to 0$, τ_n be a σ_n-a.p. for $g(t)(\sup_J \|g(t+\tau_n)-g(t)\|=\sigma_n)$ with $\sigma_n\to 0$.*

Let us now prove that $\mathscr{S}_f\subseteq\mathscr{S}_g \Rightarrow \omega_{f,g}(0+)=0$. Let $\omega_{f,g}(0+)=\rho>0$. Setting $\varepsilon=1/n$, it is possible, $\forall n$, to choose $\tau_n\neq 0$ in such a way that

$$v_f(\tau_n) \leq \frac{1}{n}, \qquad v_g(\tau_n) \geq \frac{\rho}{2}.$$

Consider now the sequence $\{s_n\}$ so defined:

$$s_{2p} = \tau_p, \qquad s_{2p-1} = 0 \qquad (p = 1,2,\cdots).$$

Then, $\forall t$,

$$\begin{aligned}\|f(t+s_{2p})-f(t+s_{2q})\| &\leq \|f(t+s_{2p})-f(t)\| + \|f(t)-f(t+s_{2q})\|\\ &\leq v_f(\tau_p)+v_f(\tau_q) \leq \frac{1}{p}+\frac{1}{q},\end{aligned}$$

$$\|f(t+s_{2p})-f(t+s_{2q-1})\| = \|f(t+s_{2p})-f(t)\| \leq v_f(\tau_p) \leq \frac{1}{p},$$

$$\|f(t+s_{2p-1})-f(t+s_{2q-1})\| = 0,$$

$$\sup_J \|g(t+s_{2p})-g(t+s_{2q-1})\| = \sup_J \|g(t+s_{2p})-g(t)\| = v_g(\tau_p) \geq \frac{\rho}{2}.$$

Therefore $\{s_n\}$ is regular for $f(t)$, but it is not regular for $g(t)$.

Let us now prove that $\omega_{f,g}(0+)=0 \Rightarrow \mathscr{S}_f\subseteq\mathscr{S}_g$. Let $\{s_n\}$ be a regular sequence for $f(t)$. Taken $\delta>0$ arbitrarily, we determine $\varepsilon_\delta>0$ in such a way that, for $0<\varepsilon\leq\varepsilon_\delta$, $\omega_{f,g}(\varepsilon)\leq\delta$. Due to the regularity of $\{s_n\}$, there exists n_δ such that, for $m,n\geq n_\delta$,

$$\sup_J \|f(t+s_n)-f(t+s_m)\| \leq \varepsilon_\delta,$$

that is,

$$v_f(s_n-s_m) = \sup_J \|f(t+s_n-s_m)-f(t)\| \leq \varepsilon_\delta.$$

Hence

$$v_g(s_n-s_m) \leq \sup_{v_f(\tau)\leq\varepsilon_\delta} v_g(\tau) = \omega_{f,g}(\varepsilon_\delta) \leq \delta,$$

that is,

$$\sup_J \|g(t+s_n)-g(t+s_m)\| \leq \delta.$$

The sequence $\{s_n\}$ is therefore regular also for $g(t)$ and the theorem is proved.

CHAPTER 2

HARMONIC ANALYSIS OF ALMOST-PERIODIC FUNCTIONS

1. THE APPROXIMATION THEOREM

The harmonic analysis of a.p. functions extends to such functions the theory of Fourier expansions of periodic functions. In the numerical case, this analysis has been made by Bohr, Weyl, De la Vallée Poussin, Bochner, who at first defined the *Fourier series* associated to a given a.p. function $f(t)$: it can then be proved (approximation theorem) that the series thus obtained is summable (by some appropriate method) to the value $f(t)$; this theory has as its starting point the *mean value theorem*. Bogoliubov has used the opposite procedure, proving at first, directly, the approximation theorem and subsequently deducing the mean value theorem and the Fourier expansion.

For a general Banach space, the first method has been generalized by Bochner [1], Kopec [2], and Zaidman [3]; the second method by Amerio [14]. In what follows, we shall keep to this second point of view, thereby extending Bogoliubov's method. It may be noted that this extension is straightforward if X is a Hilbert space. When X is a general Banach space, the proof is somewhat different from that given by Bogoliubov, because Parseval's equality for the expansion of a periodic function with values in X and (if X is not reflexive) the property of weak compactness of the sphere $\|x\| \leq 1$ do not hold. Even in the general case, it is possible however to give a rather simple direct proof of the *approximation theorem.*

Let us first observe that, $\forall a \in X$ and $\forall \lambda \in J$, the function $ae^{i\lambda t}$ is periodic. It follows that *all trigonometric polynomials*

$$P(t) = \sum_{1}^{n}{}_k\, a_k e^{i\lambda_k t} \qquad (a_k \in X,\ \lambda_k \in J)$$

are a.p.; consequently, *any function $f(t)$, which is the limit of a uniformly convergent sequence of trigonometric polynomials, is a.p.*

The fundamental *approximation theorem* enables us to establish that *all* a.p. functions can be obtained by this procedure.

I. *If $f(t)$ is a.p., there exists, $\forall \varepsilon > 0$, a trigonometric polynomial*

$$P_\varepsilon(t) = \sum_{1}^{q}{}_k\, b_k e^{i\lambda_k t} \tag{1.1}$$

such that

$$\sup_J \|f(t) - P_\varepsilon(t)\| \le \varepsilon. \tag{1.2}$$

(*a*) Let us set, for $\lambda \in J$, $T > 0$,

$$a(\lambda, T) = \frac{1}{2T}\int_{-T}^{T} f(t) e^{-i\lambda t} dt \tag{1.3}$$

and prove, at first, the following *lemma*:

$$f(t)\ a.p. \Rightarrow \mathscr{R}_{a(\lambda,T)}\ r.c..$$

We observe that the function

$$g(\lambda,t) = e^{-i\lambda t} f(t) \qquad (\lambda,\, t \in J)$$

has r.c. range; this follows from the fact that $f(t)$ is a.p. and that $e^{-i\lambda t}$ is bounded for λ, $t \in J$.

Let G be the convex extension of $g(\lambda, T)$, that is, the set of points

$$x = \sum_{1}^{n}{}_k\, \rho_k g(\lambda_k, t_k),$$

with

$$\lambda_k,\, t_k \in J, \qquad \rho_k \ge 0, \qquad \sum_{1}^{n}{}_k\, \rho_k = 1.$$

By a theorem of Mazur, the set G is also r.c., that is, the closure $\bar{G}$ is compact.

Let us prove that, $\forall \lambda \in J$, $T > 0$,

$$a(\lambda, T) \in \bar{G}.$$

We divide the interval $-T \vdash\!\dashv T$ in n equal parts of length $2T/n$ (by means of the points $-T = t_0 < t_1 < \cdots < t_n = T$) and consider the sum

$$a_n(\lambda, T) = \sum_{1}^{n}{}_k\, \frac{1}{n} f(t_k) e^{-i\lambda t_k} = \sum_{1}^{n}{}_k\, \rho_k g(\lambda, t_k)$$

with $\rho_k = 1/n$. We have, obviously,

$$a_n(\lambda, T) \in G$$

and, as

$$\lim_{n\to\infty} a_n(\lambda, T) = a(\lambda, T),$$

the thesis follows.

(*b*) We now prove the *approximation theorem*. As $f(t)$ is a.p., there exist, $\forall \varepsilon > 0$, two numbers $l > 0$ and δ, with $0 < 2\delta < l$, such that every interval of length l contains a subinterval of length 2δ completely constituted of ε-a.p. of $f(t)$.

Let, in fact, h be the inclusion length corresponding to the value $\varepsilon/2$ and choose $\delta > 0$ in such a way that, $\forall t'$ and t'' with $|t'' - t'| \leq \delta$,

$$\|f(t'') - f(t')\| \leq \frac{\varepsilon}{2};$$

this is possible, since $f(t)$ is u.c. Set $l = h + 2\delta$ and consider an interval $(a-\delta)\vdash\!\dashv(a+h+\delta)$, of length l and with arbitrary a. In $a\vdash\!\dashv a+h$ falls at least an $(\varepsilon/2)$-a.p., τ_a; furthermore, the interval $(\tau_a - \delta)\vdash\!\dashv(\tau_a + \delta)$, of length 2δ, is contained in $(a-\delta)\vdash\!\dashv(a+h+\delta)$.

Let $\tau \in (\tau_a - \delta)\vdash\!\dashv(\tau_a+\delta)$. Thus

$$\|f(t+\tau) - f(t)\| \leq \|f(t+\tau) - f(t+\tau_a)\| + \|f(t+\tau_a) - f(t)\| \leq \varepsilon$$

and the thesis is proved.

Let us define now, in $-\delta\vdash\!\dashv\delta$, a function $\varphi(t)$ such that

$$\varphi(t) \geq 0, \qquad \int_{-\delta}^{\delta} \varphi(t)dt = 4l, \qquad \int_{-\delta}^{\delta} \varphi^2(t)dt < +\infty, \tag{1.4}$$

and fix, in each interval $kl\vdash\!\dashv(k+1)l$ $(k = 0, \pm 1, \cdots)$, an open subinterval $\Delta_k = (t_k - \delta)\overline{}(t_k + \delta)$ constituted completely of ε-a.p. of $f(t)$. Moreover, set

$$\omega(t) = \begin{cases} \varphi(t - t_k) & \text{for } t \in \Delta_k \\ 0 & \text{for } t \notin \bigcup_k \Delta_k. \end{cases} \tag{1.5}$$

We now define, for $n = 1, 2, \cdots$, a periodic function $\omega_n(t)$, with period $8nl$, setting

$$\omega_n(t) = \begin{cases} \omega(t) & \text{for } |t| < nl \\ 0 & \text{for } nl \leq |t| \leq 4nl. \end{cases} \tag{1.6}$$

From (1.4), (1.5), (1.6), it follows that

$$\omega_n(t) \geq 0, \tag{1.7}$$

$$\frac{1}{8nl}\int_{-4nl}^{4nl} \omega_n(t)dt = \frac{1}{8nl}\int_{-nl}^{nl} \omega(t)dt = \frac{1}{8nl}\sum_{-n}^{n-1}{}_k \int_{\Delta_k} \varphi(t - t_k)dt = \frac{1}{4l}\int_{-\delta}^{\delta} \varphi(t)dt = 1, \tag{1.8}$$

$$\frac{1}{8nl}\int_{-4nl}^{4nl} \omega_n^2(t)dt = \frac{1}{4l}\int_{-\delta}^{\delta} \varphi^2(t)dt = \rho < +\infty. \tag{1.9}$$

We now consider the function

$$(1.10)\quad f_n(t) = \frac{1}{(8nl)^3}\int_{-nl}^{nl}\omega(\xi)d\xi\int_{-nl}^{nl}\omega(\eta)d\eta\int_{-nl}^{nl}\omega(\zeta)f(t+\xi+\eta+\zeta)d\zeta$$

and observe that the condition $\omega(\xi)\omega(\eta)\omega(\zeta)\neq 0$ implies that ξ, η, ζ are ε-a.p.; hence,

$$(1.11)\qquad \sup_J \|f(t+\xi+\eta+\zeta)-f(t)\| \leq 3\varepsilon.$$

From (1.7), (1.8), (1.10), (1.11), it follows that

$$(1.12)\quad \|f_n(t)-f(t)\| = \left\|\frac{1}{(8nl)^3}\int_{-nl}^{nl}\omega(\xi)d\xi\int_{-nl}^{nl}\omega(\eta)d\eta \times \int_{-nl}^{nl}\omega(\zeta)(f(t+\xi+\eta+\zeta)-f(t))d\zeta\right\| \leq 3\varepsilon \qquad (\forall t\in J).$$

Let us prove that

$$(1.13)\quad f_n(t) = \frac{1}{(8nl)^3}\int_{-4nl}^{4nl}\omega_n(\xi)d\xi\int_{-4nl}^{4nl}\omega_n(\eta)d\eta \times \int_{-4nl}^{4nl}\omega_n(\tau-t-\xi-\eta)f(\tau)d\tau,$$

for $|t|\leq nl$.

Observe, first of all, that, for

$$(1.14)\qquad |t|,\ |\xi|,\ |\eta| \leq nl,$$

we have

$$\int_{-nl}^{nl}\omega(\zeta)f(t+\xi+\eta+\zeta)d\zeta = \int_{-nl}^{nl}\omega_n(\zeta)f(t+\xi+\eta+\zeta)d\zeta$$
$$= \int_{-nl+t+\xi+\eta}^{nl+t+\xi+\eta}\omega_n(\tau-t-\xi-\eta)f(\tau)d\tau$$
$$= \int_{-4nl}^{4nl}\omega_n(\tau-t-\xi-\eta)f(\tau)\,d\tau.$$

In fact, $\omega_n(\zeta)=0$ for $nl\leq\zeta\leq 7nl$, that is, $\omega_n(\tau-t-\xi-\eta)=0$ for $nl+t+\xi+\eta\leq\tau\leq 7nl+t+\xi+\eta$ and, if (1.14) holds,

$$nl+t+\xi+\eta \leq 4nl \leq 7nl+t+\xi+\eta;$$

in the same way it can be seen that

$$\omega_n(\tau-t-\xi-\eta) = 0 \quad\text{for}\quad -4nl\leq\tau\leq -nl+t+\xi+\eta.$$

From (1.10) it follows then that

$$(1.15)\quad f_n(t) = \frac{1}{(8nl)^3}\int_{-nl}^{nl}\omega(\xi)d\xi\int_{-nl}^{nl}\omega(\eta)d\eta \times \int_{-4nl}^{4nl}\omega_n(\tau-t-\xi-\eta)f(\tau)d\tau \qquad (|t|\le nl).$$

Hence, by (1.6), (1.13) is proved.

Consider now the Fourier expansion (convergent in the mean on every bounded interval) of the periodic function $\omega_n(t)$:

$$(1.16)\qquad \omega_n(t) = \sum_{-\infty}^{\infty}{}_k\, \omega_k^{(n)} e^{i\nu_k^{(n)}t} \qquad \left(\nu_k^{(n)} = \frac{k\pi}{4nl}\right)$$

and observe that (owing to the orthogonality, in $-4nl \vdash\!\dashv 4nl$ of the system $\{e^{i\nu_k^{(n)}t}\}$) the sum of the series on the right-hand side of (1.16) does not depend on the order of the terms.

As, $\forall n$, $\lim_{|k|\to\infty}\omega_k^{(n)}=0$, we can arrange, $\forall n$, the Fourier coefficients $\omega_k^{(n)}$ in order of decreasing absolute values, thus obtaining a new sequence:

$$\{\theta_m^{(n)}\},\quad \text{with } \theta_m^{(n)} = \omega_{k_{m,n}}^{(n)} \qquad (m = 1,2,\cdots)$$

for which

$$(1.17)\qquad |\theta_1^{(n)}| \ge |\theta_2^{(n)}| \ge |\theta_3^{(n)}| \ge \cdots.$$

If we set

$$\nu_{k_{m,n}}^{(n)} = -\lambda_m^{(n)},$$

(1.16) becomes

$$(1.18)\qquad \omega_n(t) = \sum_{1}^{\infty}{}_m\, \theta_m^{(n)} e^{-i\lambda_m^{(n)}t},$$

this series being, as the preceding one, convergent in the mean.

From (1.8) and (1.9) it follows, by Parseval's equality, that

$$(1.19)\qquad \sum_{1}^{\infty}{}_m\, |\theta_m^{(n)}|^2 = \rho$$

and, furthermore, that

$$(1.20)\qquad \theta_m^{(n)} = \frac{1}{8nl}\int_{-4nl}^{4nl}\omega_n(t)e^{i\lambda_m^{(n)}t}dt = \frac{1}{8nl}\int_{-nl}^{nl}\omega(t)e^{i\lambda_m^{(n)}t}dt$$

$$= \frac{1}{8nl}\sum_{-n}^{n-1}{}_k\, e^{i\lambda_m^{(n)}t_k}\int_{\Delta_k}\varphi(t-t_k)e^{i\lambda_m^{(n)}(t-t_k)}dt$$

$$= \frac{1}{8nl}\left(\sum_{-n}^{n-1}{}_k\, e^{i\lambda_m^{(n)}t_k}\right)\int_{-\delta}^{\delta}\varphi(t)e^{i\lambda_m^{(n)}t}dt.$$

Hence,

$$|\theta_m^{(n)}| \leq \frac{1}{4l}\left|\int_{-\delta}^{\delta} \varphi(t)e^{i\lambda_m^{(n)}t}dt\right| \tag{1.21}$$

and, by the second of (1.4),

$$|\theta_m^{(n)}| \leq 1. \tag{1.22}$$

By (1.18) and (1.20), (1.13) can be written

$$\begin{aligned} f_n(t) &= \frac{1}{(8nl)^3}\int_{-4nl}^{4nl} \omega_n(\xi)d\xi \int_{-4nl}^{4nl} \omega_n(\eta)d\eta \\ &\qquad \times \int_{-4nl}^{4nl} f(\tau)\sum_{1}^{\infty}{}_m\, \theta_m^{(n)}e^{i\lambda_m^{(n)}(t+\xi+\eta-\tau)}d\tau \\ &= \sum_{1}^{\infty}{}_m\, (\theta_m^{(n)})^3 c_m^{(n)} e^{i\lambda_m^{(n)}t} \qquad (|t| \leq nl) \end{aligned} \tag{1.23}$$

where, by (1.3),

$$c_m^{(n)} = \frac{1}{8nl}\int_{-4nl}^{4nl} f(\tau)e^{-i\lambda_m^{(n)}\tau}d\tau = a(\lambda_m^{(n)},4nl). \tag{1.24}$$

Also

$$\|c_m^{(n)}\| \leq M, \tag{1.25}$$

where, because of the almost-periodicity,

$$\sup_J \|f(t)\| = M < +\infty.$$

Let us prove that the series

$$\sum_{1}^{\infty}{}_m\, |\theta_m^{(n)}|^3 \tag{1.26}$$

converge uniformly with respect to the index $n=1, 2,\cdots$. From (1.17) and (1.19), for $p=1, 2,\cdots$, it follows, in fact, that

$$p|\theta_p^{(n)}|^2 \leq \sum_{1}^{p}{}_m\, |\theta_m^{(n)}|^2 \leq \rho,$$

that is,

$$|\theta_p^{(n)}|^3 \leq \left(\frac{\rho}{p}\right)^{3/2}, \tag{1.27}$$

which proves the uniform convergence.

By (1.22), it is possible to select a subsequence $\{n_j\}\subset\{n\}$ such that, $\forall m$,

$$\lim_{j\to\infty} \theta_m^{(n_j)} = \theta_m, \tag{1.28}$$

and, by (1.17) and (1.27),

$$|\theta_1| \geq |\theta_2| \geq \cdots \geq |\theta_m| \geq \cdots, \tag{1.29}$$

$$|\theta_m|^3 \leq \left(\frac{\rho}{m}\right)^{3/2}. \tag{1.30}$$

By (1.29), if $\theta_m = 0$, then $\theta_{m+k} = 0$ $(k = 1, 2, \cdots)$.

We observe also that, if $\theta_m \neq 0$, the sequence $\{\lambda_m^{(n_j)}\}$ is necessarily bounded: $|\lambda_m^{(n_j)}| \leq \gamma_m < +\infty$. By a known theorem, we have, in fact,

$$\lim_{|\lambda| \to \infty} \int_{-\delta}^{\delta} \varphi(t) e^{i\lambda t} dt = 0;$$

hence, if the sequence $\{\lambda_m^{(n_j)}\}$ is not bounded, from (1.21) it follows that $\theta_m = 0$.

As $a(\lambda, T)$ has r.c. range, it is possible to select from $\{n_j\}$ a subsequence (which will again be called $\{n_j\}$) such that

$$\lim_{j \to \infty} \lambda_m^{(n_j)} = \lambda_m \qquad \text{if} \quad \theta_m \neq 0, \tag{1.31}$$

and, $\forall m$,

$$\lim_{j \to \infty} c_m^{(n_j)} = c_m, \tag{1.32}$$

where, by (1.25),

$$\|c_m\| \leq M. \tag{1.33}$$

If $\theta_m \neq 0$, it is therefore, $\forall t \in J$,

$$\lim_{j \to \infty} (\theta_m^{(n_j)})^3 c_m^{(n_j)} e^{i\lambda_m^{(n_j)} t} = \theta_m^3 c_m e^{i\lambda_m t}. \tag{1.34}$$

It is clear that relation (1.34) holds also for $\theta_m = 0$, giving an arbitrary value to λ_m.

Fix now $t \in J$ and assume that $n_j l \geq |t|$.

By (1.25) and the uniform convergence of the series (1.26), from (1.23) and (1.34) it follows that

$$\begin{aligned} \lim_{j \to \infty} f_{n_j}(t) &= \lim_{j \to \infty} \sum_{1}^{\infty}{}_m \, (\theta_m^{(n_j)})^3 c_m^{(n_j)} e^{i\lambda_m^{(n_j)} t} \\ &= \sum_{1}^{\infty}{}_m \, \theta_m^3 c_m e^{i\lambda_m t} = g(t), \end{aligned} \tag{1.35}$$

where the series that defines $g(t)$ converges, by (1.30) and (1.33), uniformly (on J).

Furthermore, by (1.12) and (1.35), we have

$$\|f(t) - g(t)\| \leq 3\varepsilon \qquad (\forall t \in J). \tag{1.36}$$

Owing to the uniform convergence, we can, therefore, take an index q such that

$$\sup_J \left\| \sum_{q+1}^{\infty} {}_m\, \theta_m^3 c_m e^{i\lambda_m t} \right\| \leq \varepsilon. \tag{1.37}$$

Hence, by (1.35), (1.36), and (1.37):

$$\sup_J \left\| f(t) - \sum_{1}^{q} {}_m\, \theta_m^3 c_m e^{i\lambda_m t} \right\| \leq 4\varepsilon, \tag{1.38}$$

which proves the theorem.

2. THE MEAN VALUE THEOREM, BOHR TRANSFORM, FOURIER SERIES, AND UNIQUENESS THEOREM

II. *If $f(t)$ is a.p., the mean value*

$$\mathscr{M}(f(t)) = \lim_{T\to\infty} \frac{1}{2T} \int_{-T}^{T} f(t)dt \tag{2.1}$$

exists; it is, furthermore,

$$\lim_{T\to\infty} \frac{1}{2T} \int_{-T+s}^{T+s} f(t)dt = \mathscr{M}(f(t)) \tag{2.2}$$

uniformly with respect to $s \in J$.

Setting

$$\psi(\lambda) = \begin{cases} 1 & \text{for } \lambda = 0 \\ 0 & \text{for } \lambda \neq 0 \end{cases} \tag{2.3}$$

and since, for $T > 0$, $s \in J$,

$$\frac{1}{2T} \int_{-T+s}^{T+s} e^{i\lambda t} dt = \begin{cases} 1 & \text{for } \lambda = 0 \\ e^{i\lambda s} \dfrac{\sin \lambda T}{\lambda T} & \text{for } \lambda \neq 0, \end{cases}$$

we obtain, $\forall \lambda \in J$,

$$\lim_{T\to\infty} \frac{1}{2T} \int_{-T+s}^{T+s} e^{i\lambda t} dt = \mathscr{M}(e^{i\lambda t}) = \psi(\lambda) \tag{2.4}$$

uniformly with respect to s.

From (2.3) and (2.4) it follows that (2.2) holds for an arbitrary trigonometric polynomial

$$P(t) = \sum_{1}^{n} {}_k\, a_k e^{i\lambda_k t},$$

and we have

$$\mathscr{M}(P(t)) = \sum_{1}^{n} {}_k\, a_k \psi(\lambda_k). \tag{2.5}$$

Relation (2.2) then follows, in the general case, from the approximation theorem. Taken $\varepsilon > 0$ arbitrarily, let in fact $P_\varepsilon(t)$ be a trigonometric polynomial such that

$$\sup_J \|f(t) - P_\varepsilon(t)\| \leq \varepsilon. \tag{2.6}$$

If

$$\mathscr{M}(f(t);\, T,s) = \frac{1}{2T}\int_{-T+s}^{T+s} f(t)dt,$$

it is then

$$\begin{aligned}
&\|\mathscr{M}(f(t);\, T',s) - \mathscr{M}(f(t);\, T'',s)\| \\
&\qquad \leq \|\mathscr{M}(f(t)-P_\varepsilon(t);\, T',s)\| + \|\mathscr{M}(P_\varepsilon(t);\, T',s) - \mathscr{M}(P_\varepsilon(t);\, T'',s)\| \\
&\qquad\quad + \|\mathscr{M}(P_\varepsilon(t)-f(t);\, T'',s)\| \\
&\qquad \leq \|\mathscr{M}(P_\varepsilon(t);\, T',s) - \mathscr{M}(P_\varepsilon(t);\, T'',s)\| + 2\varepsilon
\end{aligned}$$

and further, for T', $T'' \geq T_\varepsilon$ and $\forall s \in J$,

$$\|\mathscr{M}(f(t);\, T',s) - \mathscr{M}(f(t);\, T'',s)\| \leq 3\varepsilon.$$

Relation (2.2) is therefore proved.

Let us now observe that $f(t)$ a.p. $\Rightarrow$ $f(t)e^{-i\lambda t}$ a.p., $\forall \lambda \in J$. The function of λ:

$$a(\lambda; f(t)) = \mathscr{M}(f(t)e^{-i\lambda t}), \tag{2.7}$$

is obviously defined from J to X. We shall call $a(\lambda; f(t))$ the *Bohr transform of the a.p. function $f(t)$.*

Setting

$$\sup_J \|f(t)\| = M, \tag{2.8}$$

it follows immediately from (2.7) that

$$\|a(\lambda; f(t))\| \leq M.$$

III. $a(\lambda; f(t)) = 0$ *on the whole of J, with, at most, the exception of a sequence* $\{\lambda_n\}$.

Let, in fact, $\{P_k(t)\}$ be a sequence of approximating trigonometric polynomials, such that

$$\sup_J \|f(t) - P_k(t)\| \leq \frac{1}{k} \qquad (k = 1, 2, \cdots).$$

Setting

$$P_k(t) = \sum_{1}^{n_k}{}_m\, a_{km}e^{i\lambda_{km}t}, \qquad \{\mu_n\} = \bigcup_{k,m} \lambda_{km},$$

we have, for $\lambda \notin \{\mu_n\}$ (observing that, in this case, by (2.5), $a(\lambda;\ P_k(t))=0$, $\forall k$),

$$\|a(\lambda; f(t))\| = \|a(\lambda;\ P_k(t))+a(\lambda; f(t)-P_k(t))\|$$
$$= \|a(\lambda; f(t)-P_k(t))\| \leq \frac{1}{k}.$$

Hence $a(\lambda; f(t))=0$.

The values $\{\lambda_n\}$ for which $a(\lambda_n; f(t))\neq 0$ are called the *characteristic exponents of* $f(t)$ and the sequence $\{\lambda_n\}$ constitutes the *spectrum* of $f(t)$; obviously, $\{\lambda_n\}\subseteq\{\mu_n\}$ and setting

$$a_n = a(\lambda_n; f(t))$$

we can associate (for the time being in a purely formal way) to $f(t)$ the Fourier series

$$f(t) \sim \sum_{1}^{\infty}{}_n\, a_n e^{i\lambda_n t},$$

thus obtaining, for $f(t)$, a *decomposition formula* which constitutes its *harmonic analysis*. The a_n's are called the *Fourier coefficients* of $f(t)$.

Setting

$$a(\lambda,T; f(t)) = \frac{1}{2T}\int_{-T}^{T} f(t)e^{-i\lambda t}dt, \tag{2.9}$$

let us prove the following theorem.

IV. $a(\lambda'; f(t))=0 \Rightarrow \lim_{\lambda\to\lambda',T\to\infty} a(\lambda,T; f(t))=0.$

The Bohr transform is therefore continuous at all points at which it vanishes. Furthermore, we have

$$\lim_{\substack{\lambda\to\infty\\ T\to\infty}} a(\lambda,T; f(t)) = 0.$$

Let $P(t)$ be a trigonometric polynomial such that

$$\sup_J \|f(t)-P(t)\| \leq \varepsilon.$$

We have then

$$P(t) = \sum_{1}^{n}{}_k\, b_k e^{i\nu_k t}+b'e^{i\lambda' t} \qquad (\nu_k \neq \lambda'),$$

and

$$\|b'\| = \|\mathscr{M}(P(t)e^{-i\lambda' t})\| \leq \|\mathscr{M}(P(t)-f(t))e^{-i\lambda' t}\|$$
$$+\|\mathscr{M}(f(t)e^{-i\lambda' t})\| \leq \varepsilon.$$

Therefore, setting $Q(t)=P(t)-b'e^{i\lambda' t}$, we obtain

$$\sup_J \|f(t)-Q(t)\| \leq 2\varepsilon.$$

It is then possible to define a sequence of trigonometric polynomials such that

$$\sup_J \|P_n(t)-f(t)\| \le \frac{1}{n} \tag{2.10}$$

and such that λ' is not a characteristic exponent for $P_n(t)$, $\forall n$.

Let us prove that

$$\lim_{\substack{\lambda\to\lambda' \\ T\to\infty}} a(\lambda,T;\, P_n(t)) = 0, \qquad \forall n. \tag{2.11}$$

Setting

$$P_n(t) = \sum_{1}^{p_n}{}_k\, b_{nk}e^{i\lambda_{nk}t},$$

we have, for $\lambda \neq \lambda_{n1},\cdots,\lambda_{np_n}$,

$$a(\lambda,T;\, P_n(t)) = \frac{1}{T}\sum_{1}^{p_n}{}_k\, b_{nk}\frac{\sin(\lambda-\lambda_{nk})T}{(\lambda-\lambda_{nk})}.$$

Relation (2.11) then follows, being $\lambda' \neq \lambda_{nk}$ $(k=1,\cdots,p_n)$. Furthermore,

$$\begin{aligned} \|a(\lambda,T;f(t))\| &\le \|a(\lambda,T;\, P_n(t))\| + \|a(\lambda,T;f(t)-P_n(t))\| \\ &\le \|a(\lambda,T;\, P_n(t))\| + \frac{1}{n}. \end{aligned} \tag{2.12}$$

Having fixed $\varepsilon>0$ arbitrarily, let $n_\varepsilon \ge 1/\varepsilon$; then

$$\|a(\lambda,T;f(t))\| \le \|a(\lambda,T;\, P_{n_\varepsilon}(t))\| + \varepsilon. \tag{2.13}$$

If $0<|\lambda-\lambda'|\le\delta_\varepsilon$, $T\ge T_\varepsilon$, it follows from (2.11) that

$$\|a(\lambda,T;\, P_{n_\varepsilon}(t))\| \le \varepsilon$$

and, by (2.13),

$$\|a(\lambda,T;f(t))\| \le 2\varepsilon,$$

which proves the theorem. In an analogous way, it can be shown that

$$\lim_{\substack{\lambda\to\infty \\ T\to\infty}} a(\lambda,T;f(t)) = 0.$$

V. *The characteristic exponents of the trigonometric polynomial $P_\varepsilon(t)$ defined by* (1.38) *in the approximation theorem, belong,* $\forall\varepsilon>0$, *to the sequence* $\{\lambda_n\}$ *of the characteristic exponents of* $f(t)$.

Let, in fact, $a(\lambda';f(t))=0$ and assume that $\lambda'=\mu_k$, characteristic exponent of the polynomial $P_\varepsilon(t)$, where we have set, by (1.38),

$$P_\varepsilon(t) = \sum_{1}^{q}{}_m\, \theta_m^3 c_m e^{i\mu_m t} \qquad (c_m \neq 0).$$

By (1.24), (1.31), (1.32), and by theorem IV, we deduce

$$c_k = a(\mu_k; f(t)) = a(\lambda', f(t)) = 0, \tag{2.14}$$

which is absurd. Hence it must be $a(\mu_k; f(t)) \neq 0 \Rightarrow \mu_k \in \{\lambda_n\}$.

VI (*Uniqueness Theorem*). *$f(t)$ and $g(t)$ X-a.p., and*

$$a(\lambda; f(t)) = a(\lambda; g(t)) \Rightarrow f(t) = g(t).$$

In fact, all the Fourier coefficients of the function $f(t)-g(t)$ vanish. By (2.14), the corresponding approximation polynomials considered in theorem V are identically zero, $\forall \varepsilon > 0$. It follows that $f(t) \equiv g(t)$.

Hence, *there exists a one-to-one correspondence between a.p. functions and their Bohr transforms.*

VII. $f(t)$ *a.p.* $\Rightarrow \lim_{n\to\infty} a_n = 0$.

Let, in fact, $P_\varepsilon(t)$ be an approximation polynomial, such that

$$\sup_J \|f(t) - P_\varepsilon(t)\| \leq \varepsilon$$

and let us fix n_ε in such a way that, for $n > n_\varepsilon$, the characteristic exponents of $f(t)$ are different from those of $P_\varepsilon(t)$.

Hence $\mathscr{M}(P_\varepsilon(t)e^{-i\lambda_n t}) = 0$ and, consequently,

$$\|a_n\| = \|\mathscr{M}(f(t)e^{-i\lambda_n t})\| = \|\mathscr{M}(f(t) - P_\varepsilon(t))e^{-i\lambda_n t}\| \leq \varepsilon.$$

3. BOCHNER'S APPROXIMATION POLYNOMIAL

To Bochner we owe the definition of an important approximation polynomial, which is suggested by Fejer's classical theorem on the Cesaro-summability of the Fourier series of a periodic function.

First, let $B = \{\beta_1, \beta_2, \cdots, \beta_n, \cdots\}$ be a *basis* for the sequence $\{\lambda_n\}$ of the characteristic exponents of the function $f(t)$; this means that the real numbers β_n are *linearly independent* (i.e., $\sum_{(1)k}^{(n)} c_k\beta_k = 0$, c_k integers, $\Rightarrow c_1 = \cdots = c_n = 0$) and that *each λ_n is a linear combination with rational coefficients of a finite number of the β_k's.*

It is clear that a basis always exists: if the λ_n's are linearly independent, we may take $\beta_n = \lambda_n$; otherwise B may be obtained by eliminating successively those λ_n's that are linear combinations of the preceding ones. Denoting by $\tilde{B}$ the sequence of all finite linear combinations, with rational coefficients, of the β_k's, we will therefore have $\{\lambda_n\} \subseteq \tilde{B}$.

Let us now denote by $\Pi_n(t)$ *Fejer's kernel*:

$$\Pi_n(t) = \sum_{-n}^{n}{}_{\nu} \left(1 - \frac{|\nu|}{n}\right) e^{-i\nu t} = \frac{1}{n}\left(\frac{\sin nt/2}{\sin t/2}\right)^2 \tag{3.1}$$

and let $\alpha_1, \cdots, \alpha_k$ be k linearly independent numbers.

If $f(t)$ is a.p., we shall call *Bochner's polynomial* the trigonometric polynomial:

$$(3.2)\quad Q^{\alpha_1\cdots\alpha_k}_{n_1\cdots n_k}(f(t)) = \mathscr{M}_s\{f(s+t)\Pi_{n_1}(\alpha_1 s)\cdots\Pi_{n_k}(\alpha_k s)\}$$

$$= \lim_{T\to\infty}\frac{1}{2T}\int_{-T}^{T} f(s+t)$$

$$\times\left[\sum_{-n_1}^{n_1}{}_{\nu_1}\cdots\sum_{-n_k}^{n_k}{}_{\nu_k}\left(1-\frac{|\nu_1|}{n_1}\right)\cdots\left(1-\frac{|\nu_k|}{n_k}\right)\right]$$

$$\times\exp\left(-i\sum_{1}^{k}{}_{j}\,\nu_j\alpha_j s\right)ds$$

$$= \sum_{-n_1}^{n_1}{}_{\nu_1}\cdots\sum_{-n_k}^{n_k}{}_{\nu_k}\left(1-\frac{|\nu_1|}{n_1}\right)\cdots\left(1-\frac{|\nu_k|}{n_k}\right)$$

$$\times a\left(\sum_{1}^{k}{}_{j}\,\nu_j\alpha_j; f(s)\right)\exp\left(i\sum_{1}^{k}{}_{j}\,\nu_j\alpha_j t\right).$$

The kernels possess the following properties:

$$(3.3)\qquad \Pi_n(s)\geq 0,\qquad \mathscr{M}(\Pi_{n_1}(\alpha_1 s)\cdots\Pi_{n_k}(\alpha_k s)) = 1.$$

From these and (3.2) follows then the inequality

$$(3.4)\qquad \|Q^{\alpha_1\cdots\alpha_k}_{n_1\cdots n_k}(f(t))\| \leq \sup_J \|f(t)\|.$$

Let us now set, if the basis B is infinite,

$$(3.5)\quad Q_m(f(t)) = Q^{\beta_1/m!\cdots\beta_m/m!}_{(m!)^2\cdots(m!)^2}(f(t))$$

$$= \sum_{-(m!)^2}^{(m!)^2}{}_{\nu_1}\cdots\sum_{-(m!)^2}^{(m!)^2}{}_{\nu_m}\left(1-\frac{|\nu_1|}{(m!)^2}\right)\cdots\left(1-\frac{|\nu_m|}{(m!)^2}\right)$$

$$\times a\left(\sum_{1}^{m}{}_{j}\,\nu_j\frac{\beta_j}{m!}; f(s)\right)\exp\left(i\sum_{1}^{m}{}_{j}\,\nu_j\frac{\beta_j}{m!}t\right).$$

If the basis B is finite ($B=\{\beta_1,\cdots,\beta_k\}$), we shall set, for $m\geq k$:

$$(3.6)\quad Q_m(f(t)) = Q^{\beta_1/m!\cdots\beta_k/m!}_{(m!)^2\cdots(m!)^2}(f(t))$$

$$= \sum_{-(m!)^2}^{(m!)^2}{}_{\nu_1}\cdots\sum_{-(m!)^2}^{(m!)^2}{}_{\nu_k}\left(1-\frac{|\nu_1|}{(m!)^2}\right)\cdots\left(1-\frac{|\nu_k|}{(m!)^2}\right)$$

$$\times a\left(\sum_{1}^{k}{}_{j}\,\nu_j\frac{\beta_j}{m!}; f(s)\right)\exp\left(i\sum_{1}^{k}{}_{j}\,\nu_j\frac{\beta_j}{m!}t\right).$$

We shall prove that

$$(3.7)\qquad \lim_{m\to\infty} Q_m(f(t)) = f(t)$$

uniformly.

Let us observe, first of all, that, for linear combinations of the type

$$f(t) = \sum_1^n {}_k \gamma_k f_k(t), \qquad g(t) = \sum_1^n {}_k c_k \psi_k(t)$$

(where γ_k, $\psi_k(t)$ have complex values and c_k, $f_k(t)$ take their values in X),

$$(3.8) \qquad \begin{aligned} Q_m\left(\sum_1^n {}_k \gamma_k f_k(t)\right) &= \sum_1^n {}_k \gamma_k Q_m(f_k(t)), \\ Q_m\left(\sum_1^n {}_k c_k \psi_k(t)\right) &= \sum_1^n {}_k c_k Q_m(\psi_k(t)). \end{aligned}$$

Now let $\psi(t) = e^{i\lambda' t}$, $\lambda' \in \tilde{B}$, so that

$$(3.9) \qquad \lambda' = \sum_1^l {}_k r_k \beta_k,$$

with $r_k = (p_k/q_k)$ rational numbers and p_k, q_k relatively prime integers.

It is then

$$(3.10) \qquad a\left(\sum_1^m {}_k \frac{\nu_k}{m!} \beta_k; e^{i\lambda' s}\right) = \begin{cases} 1 & \text{for } \lambda' = \sum_1^m {}_k \dfrac{\nu_k}{m!} \beta_k \\ 0 & \text{for } \lambda' \neq \sum_1^m {}_k \dfrac{\nu_k}{m!} \beta_k. \end{cases}$$

Let us choose m in such a way that

$$(3.11) \qquad m > l;\ m! \geq |r_1|, \cdots, |r_l|;\ m! \ \text{ divisible by } \ q_1, \cdots, q_l.$$

In this case the first equation of (3.10) is verified (being the β_k's linearly independent) only when

$$(3.12) \qquad \begin{aligned} &\frac{\nu_k}{m!} = r_k, \quad \text{that is,} \quad \nu_k = p_k \frac{m!}{q_k} \qquad (k = 1, \cdots, l), \\ &\nu_k = 0 \qquad (k > l). \end{aligned}$$

We obtain in this way one and only one m-ple of integers and, furthermore,

$$|\nu_k| = |r_k| m! \leq (m!)^2.$$

By (3.5), (3.10), and (3.12), we have then

$$Q_m(e^{i\lambda' t}) = \left(1 - \frac{|r_1|}{m!}\right) \cdots \left(1 - \frac{|r_l|}{m!}\right) e^{i\lambda' t}$$

and, consequently,

$$(3.13) \qquad \lim_{m \to \infty} Q_m(e^{i\lambda' t}) = e^{i\lambda' t}$$

uniformly.

Let

$$P(t) = \sum_{1}^{n}{}_{k}\, c_k e^{i\lambda_k t}$$

be an arbitrary trigonometric polynomial, from J to X, and such that $\{\lambda_k\} \subseteq \tilde{B}$.

By the second of (3.8) and by (3.13), it follows that

$$\lim_{m\to\infty} Q_m(P(t)) = \sum_{1}^{n}{}_{k}\, c_k e^{i\lambda_k t} = P(t), \tag{3.14}$$

uniformly.

Let $f(t)$ be an arbitrary a.p. function. If $\{\lambda_n\}$ is the sequence of characteristic exponents and B is a basis such that $\{\lambda_n\} \subseteq \tilde{B}$, we can define (by theorem V) a sequence of trigonometric polynomials, of the form

$$P_n(t) = \sum_{1}^{p_n}{}_{k}\, a_{nk} e^{i\lambda_k t},$$

such that

$$\sup_J \|f(t) - P_n(t)\| \leq \frac{1}{n}. \tag{3.15}$$

Hence, by (3.4),

$$\sup_J \|Q_m(f(t) - P_n(t))\| \leq \frac{1}{n}. \tag{3.16}$$

By (3.15) and (3.16) it follows that:

$$\begin{aligned} \|f(t) - Q_m(f(t))\| &\leq \|f(t) - P_n(t)\| + \|P_n(t) - Q_m(P_n(t))\| \\ &\quad + \|Q_m(P_n(t) - f(t))\| \\ &\leq \frac{2}{n} + \|P_n(t) - Q_m(P_n(t))\|. \end{aligned}$$

Taken $\varepsilon > 0$ arbitrarily, let n_ε be such that

$$\frac{2}{n_\varepsilon} \leq \varepsilon.$$

Moreover, for $m \geq m_\varepsilon$, by (3.14),

$$\|P_{n_\varepsilon}(t) - Q_m(P_{n_\varepsilon}(t))\| \leq \varepsilon$$

and (3.7) is proved.

Observation. In the expression of the polynomial $Q_m(t)$, given by (3.5), we can extend the summation only to those terms for which

$$a\left(\sum_{1}^{m}{}_{j}\, \nu_j \frac{\beta_j}{m!}; f(s)\right) \neq 0,$$

that is,

$$\sum_{1}^{m}{}_{j}\, \nu_j \frac{\beta_j}{m!} \in \{\lambda_n\}.$$

We obtain, in this way,

$$Q_m(f(t)) = \sum_{1}^{r_m}{}_k \mu_{mk} a_k e^{i\lambda_k t} \tag{3.17}$$

where $a_k = a(\lambda_k; f(s))$ and the *convergence factors* μ_{mk} *depend only on the sequence of characteristic exponents* $\{\lambda_n\}$ *and satisfy the inequalities*

$$0 < \mu_{mk} \leq 1.$$

4. THE CASE OF HILBERT SPACE. PARSEVAL'S EQUATION

VIII. *X Hilbert space* $\Rightarrow \mathcal{M}(\|f(t)\|^2) = \sum_{(1)}^{(\infty)}{}_n \|a_n\|^2$.

Hence *Parseval's equation*, which is valid for the Fourier expansion of a periodic function, *holds also for a.p. functions.*

For the proof, if $\{\lambda_k\}$ is the sequence of characteristic exponents of $f(t)$, we consider, at first, all polynomials of order n, n fixed, of the type

$$P(t) = \sum_{1}^{n}{}_k b_k e^{i\lambda_k t}.$$

Setting

$$\mathcal{M}(\|f(t) - P(t)\|^2) = \Phi(b_1, \cdots, b_n), \tag{4.1}$$

let us prove that *there exists one, and only one, polynomial* $\tilde{P}(t)$ *in correspondence to which the function* $\Phi(b_1, \cdots, b_n)$ *takes on its smallest value; precisely,*

$$\tilde{P}(t) = \sum_{1}^{n}{}_k a_k e^{i\lambda_k t}, \tag{4.2}$$

where a_k *is the Fourier coefficient of* $f(t)$ *corresponding to the exponent* λ_k.

In fact,

$$\begin{aligned} 0 \leq \mathcal{M}(\|f(t) - P(t)\|^2) &= \lim_{T \to \infty} \frac{1}{2T} \int_{-T}^{T} (f(t) - P(t), f(t) - P(t))dt \\ &= \mathcal{M}(\|f(t)\|^2) + \mathcal{M}(\|P(t)\|^2) - \mathcal{M}(f(t), P(t)) - \mathcal{M}(P(t), f(t)) \end{aligned} \tag{4.3}$$

and, by (4.1),

$$\mathcal{M}(\|P(t)\|^2) = \sum_{j,k}^{1..n} b_j \bar{b}_k \lim_{T \to \infty} \frac{1}{2T} \int_{-T}^{T} e^{i(\lambda_j - \lambda_k)t} dt = \sum_{1}^{n}{}_k \|b_k\|^2. \tag{4.4}$$

Moreover,

$$\begin{aligned} \mathcal{M}(f(t), P(t)) &= \sum_{1}^{n}{}_k \lim_{T \to \infty} \frac{1}{2T} \int_{-T}^{T} (f(t), b_k) e^{-i\lambda_k t} dt \\ &= \sum_{1}^{n}{}_k \lim_{T \to \infty} \left(\frac{1}{2T} \int_{-T}^{T} f(t) e^{-i\lambda_k t} dt, b_k \right) = \sum_{1}^{n}{}_k (a_k, b_k), \end{aligned} \tag{4.5}$$

$$\mathcal{M}(P(t), f(t)) = \sum_{1}^{n}{}_k (b_k, a_k). \tag{4.6}$$

From (4.1) and (4.3) through (4.6), it follows that

$$
\begin{aligned}
(4.7) \qquad 0 \leq \Phi(b_1,\cdots,b_n) &= \mathcal{M}(\|f(t)\|^2)+\sum_{1}^{n}{}_k \|b_k\|^2-\sum_{1}^{n}{}_k (a_k,b_k)-\sum_{1}^{n}{}_k (b_k,a_k) \\
&= \mathcal{M}(\|f(t)\|^2)-\sum_{1}^{n}{}_k \|a_k\|^2+\sum_{1}^{n}{}_k \|a_k-b_k\|^2.
\end{aligned}
$$

Therefore, $\Phi(b_1,\cdots,b_n)$ takes on its minimum value when $b_1=a_1,\cdots,$ $b_n = a_n$ (and only in this case). Hence, by (4.2),

$$\mathcal{M}(\|f(t)-P(t)\|^2) = \mathcal{M}(\|f(t)\|^2)-\sum_{1}^{n}{}_k \|a_k\|^2 \geq 0,$$

that is, the series $\sum_{(1)}^{(\infty)}{}_k \|a_k\|^2$ is convergent and satisfies the condition (*Bessel's inequality*)

$$(4.8) \qquad \sum_{1}^{\infty}{}_k \|a_k\|^2 \leq \mathcal{M}(\|f(t)\|^2).$$

Let us show that in (4.8) the equality sign holds; the theorem will then be proved.

Let $P_\varepsilon(t)$ be an approximation polynomial corresponding to an arbitrary $\varepsilon>0$, with characteristic exponents $\in\{\lambda_n\}$ and such that

$$\sup_J \|f(t)-P_\varepsilon(t)\| \leq \varepsilon.$$

Then

$$\mathcal{M}(\|f(t)-P_\varepsilon(t)\|^2) \leq \varepsilon^2.$$

Let n_ε be the degree of $P_\varepsilon(t)$ and

$$\tilde{P}_\varepsilon(t) = \sum_{1}^{n_\varepsilon}{}_k a_k e^{i\lambda_k t}$$

the polynomial that minimizes $\Phi(b_1,\cdots, b_{n_\varepsilon})$. Therefore,

$$
\begin{aligned}
0 \leq \mathcal{M}(\|f(t)-\tilde{P}_\varepsilon(t)\|^2) &= \mathcal{M}(\|f(t)\|^2)-\sum_{1}^{n_\varepsilon}{}_k \|a_k\|^2 \\
&\leq \mathcal{M}(\|f(t)-P_\varepsilon(t)\|^2) \leq \varepsilon^2,
\end{aligned}
$$

which proves the theorem.

5. CHARACTERISTIC EXPONENTS AND ALMOST PERIODS

We shall now make use of an important theorem of Kronecker, the proof of which can be deduced, following Bohr, from the theory of a.p. functions.

Let us consider a *system of n (congruencial) inequalities*

$$(5.1)\qquad |\lambda_k\tau-\theta_k| \le \delta \pmod{2\pi} \qquad (k=1,\cdots,n)$$

where τ is the unknown and $\lambda_1,\cdots,\lambda_n,\ \theta_1,\cdots,\theta_n,\ \delta>0$ are given real numbers.

Kronecker's theorem gives the condition for the system (5.1) to admit a solution, $\forall\delta>0$.

IX. *System* (5.1) *admits a solution* τ_δ, $\forall\delta>0$, *if and only if every time that a relation of the type*

$$(5.2)\qquad \sum_{1}^{n}{}_k\, q_k\lambda_k = 0 \qquad (q_1,\cdots,q_n \textit{ integers})$$

holds, then we have also

$$(5.3)\qquad \sum_{1}^{n}{}_k\, q_k\theta_k = 0 \qquad \pmod{2\pi}.$$

The condition is necessary. Let (5.2) be verified for certain integers $q_1,\cdots,q_n$ and let τ_δ be a solution of system (5.1), that is, of the system

$$(5.4)\qquad |\lambda_k\tau-\theta_k+2\pi m_k| \le \delta,$$

$m_1,\cdots,m_n$ being integers (depending on δ).

It follows that

$$(5.5)\qquad \left|\sum_{1}^{n}{}_k\, q_k\theta_k-2\pi\sum_{1}^{n}{}_k\, q_k m_k\right| = \left|\left(\sum_{1}^{n}{}_k\, q_k\lambda_k\right)\tau_\delta-\sum_{1}^{n}{}_k\, q_k\theta_k+2\pi\sum_{1}^{n}{}_k\, q_k m_k\right| \le \delta\sum_{1}^{n}{}_k\, |q_k|,$$

that is (setting $m_\delta=\sum_{(1)k}^{(n)} q_k m_k$),

$$-\delta\sum_{1}^{n}{}_k\, |q_k|+\sum_{1}^{n}{}_k\, q_k\theta_k \le 2\pi m_\delta \le \sum_{1}^{n}{}_k\, q_k\theta_k+\delta\sum_{1}^{n}{}_k\, |q_k|.$$

Hence, if

$$\delta\sum_{1}^{n}{}_k\, |q_k| < \pi,$$

the integer m_δ does not depend on δ; from (5.5) then follows (5.3) when $\delta\to 0$.

The condition is sufficient. Let us consider the exponential polynomials

$$(5.6)\qquad \varphi(t) = \prod_{1}^{n}{}_k\, (1+e^{i(\lambda_k t-\theta_k)}) \qquad (t\in J)$$

and

$$(5.7)\qquad \psi(t_1,\cdots,t_n) = \prod_{1}^{n}{}_k\, (1+e^{it_k}) \qquad (t_1,\cdots,t_n\in J).$$

Obviously,

$$\varphi(t) = \psi(\lambda_1 t-\theta_1,\cdots, \lambda_n t-\theta_n); \tag{5.8}$$

furthermore, $\varphi(t)$ is an a.p. function.

If p is any positive integer, then

$$\begin{aligned}(\psi(t_1,\cdots, t_n))^p &= \left(\sum_{0}^{p}{}_{q_1} \binom{p}{q_1} e^{iq_1t_1}\right)\cdots\left(\sum_{0}^{p}{}_{q_n} \binom{p}{q_n} e^{iq_nt_n}\right)\\ &= \sum_{q_1,\cdots,q_n}^{0\cdots p} \binom{p}{q_1}\cdots\binom{p}{q_n} \exp\left(i\sum_{1}^{n}{}_{k}\, q_k t_k\right)\end{aligned} \tag{5.9}$$

and

$$(\varphi(t))^p = \sum_{q_1,\cdots,q_n}^{0\cdots p} \binom{p}{q_1}\cdots\binom{p}{q_n} \exp\left[i\left(t\sum_{1}^{n}{}_{k}\, q_k\lambda_k - \sum_{1}^{n}{}_{k}\, q_k\theta_k\right)\right]. \tag{5.10}$$

We now make all possible reductions on the right-hand side of (5.10).

For this, we observe that, for certain integers $q_1',\cdots, q_n'$ and $q_1'',\cdots, q_n''$, we can have, on J,

$$\exp\left(it\sum_{1}^{n}{}_{k}\, q_k'\lambda_k\right) \equiv \exp\left(it\sum_{1}^{n}{}_{k}\, q_k''\lambda_k\right) \tag{5.11}$$

if, and only if,

$$\sum_{1}^{n}{}_{k}\, (q_k'-q_k'')\lambda_k = 0. \tag{5.12}$$

Then, by (5.2) and (5.3),

$$\sum_{1}^{n}{}_{k}\, (q_k'-q_k'')\theta_k = 0 \pmod{2\pi}$$

and, consequently,

$$\begin{aligned}&\binom{p}{q_1'}\cdots\binom{p}{q_n'} \exp\left[i\left(t\sum_{1}^{n}{}_{k}\, q_k'\lambda_k - \sum_{1}^{n}{}_{k}\, q_k'\theta_k\right)\right]\\ &\qquad + \binom{p}{q_1''}\cdots\binom{p}{q_n''} \exp\left[i\left(t\sum_{1}^{n}{}_{k}\, q_k''\lambda_k - \sum_{1}^{n}{}_{k}\, q_k''\theta_k\right)\right]\\ &= \left\{\binom{p}{q_1'}\cdots\binom{p}{q_n'} + \binom{p}{q_1''}\cdots\binom{p}{q_n''}\right\} \exp\left[i\left(t\sum_{1}^{n}{}_{k}\, q_k'\lambda_k - \sum_{1}^{n}{}_{k}\, q_k'\theta_k\right)\right].\end{aligned}$$

Observe now that the square of the sum of the absolute values of complex numbers having the same argument (mod 2π) is not smaller than the sum of the squares of the absolute values themselves. Therefore, once the reductions are effected, the sum of the squares of the absolute values of the single terms of $(\varphi(t))^p$ is not smaller than the sum of the squares of the coefficients of $(\varphi(t_1,\cdots, t_n))^p$.

By Parseval's equality, the first of these sums equals $\mathscr{M}(|\varphi(t)|^{2p})$. It follows that

$$\mathscr{M}(|\varphi(t)|^{2p}) \geq \sum_{q_1,\cdots,q_n}^{0\cdots p} \binom{p}{q_1}^2 \binom{p}{q_2}^2 \cdots \binom{p}{q_n}^2 = \left(\sum_0^p {}_q \binom{p}{q}^2\right)^n. \tag{5.13}$$

Now

$$(1+t)^{2p} = \sum_0^{2p} {}_k \binom{2p}{k} t^k = \left(\sum_0^p {}_k \binom{p}{k} t^k\right)^2$$

and therefore, equating the coefficients of t^p,

$$\binom{2p}{p} = \sum_0^p {}_k \binom{p}{k}\binom{p}{p-k} = \sum_0^p {}_k \binom{p}{k}^2. \tag{5.14}$$

Furthermore,

$$\mathscr{M}(|\varphi(t)|^{2p}) \leq (\sup_J |\varphi(t)|)^{2p} \tag{5.15}$$

and, by (5.13), (5.14) and (5.15),

$$\sup_J |\varphi(t)| \geq \binom{2p}{p}^{n/2p}.$$

Hence, by Stirling's formula,

$$\sup_J |\varphi(t)| \geq 2^n.$$

From (5.6), on the other hand,

$$\sup_J |\varphi(t)| \leq 2^n.$$

Therefore,

$$\sup_J |\varphi(t)| = 2^n.$$

For any δ, with $0 < \delta \leq \pi/2$, there exists a value τ_δ such that

$$|\varphi(\tau_\delta)| \geq 2^{n-1}|1+e^{i\delta}|. \tag{5.16}$$

For the same τ_δ we cannot have (even for a single value of k), by (5.6),

$$|1+e^{i(\lambda_k\tau_\delta-\theta_k)}| < |1+e^{i\delta}|.$$

Therefore, for any k,

$$|1+e^{i(\lambda_k\tau_\delta-\theta_k)}| \geq |1+e^{i\delta}|$$

that is,

$$|\lambda_k\tau_\delta-\theta_k| \leq \delta \pmod{2\pi}.$$

The theorem is therefore proved.

Corollary. *If system (5.1) admits, $\forall\delta>0$, a solution, then there exists a length ρ_δ such that every interval of length ρ_δ contains a solution of the system itself.*

Let us, in fact, set $\eta_\delta=2^{n-1}\{|1+e^{i(\delta/2)}|-|1+e^{i\delta}|\}$, $0<\delta<\pi$, and let σ be any η_δ-a.p. for $\varphi(t)$. It is then, $\forall t\in J$,

$$|\varphi(t+\sigma)| \geq |\varphi(t)|-\eta_\delta.$$

Moreover, by (5.16), there exists $\tau_{\delta/2}$ such that

$$|\varphi(\tau_{\delta/2})| \geq 2^{n-1}|1+e^{i(\delta/2)}|.$$

Hence,

$$|\varphi(\tau_{\delta/2}+\sigma)| \geq |\varphi(\tau_{\delta/2})|-\eta_\delta \geq 2^{n-1}|1+e^{i\delta}|,$$

from which follows, as before,

$$|\lambda_k(\tau_{\delta/2}+\sigma)-\theta_k| \leq \delta \qquad (\text{mod } 2\pi).$$

As the set $\{\tau_{\delta/2}+\sigma\}$ is r.d., denoting by ρ_δ the corresponding inclusion length, the thesis is proved.

X. *Let*

$$f(t) \sim \sum_{1}^{\infty}{}_{n}\, a_n e^{i\lambda_n t}$$

$$g(t) \sim \sum_{1}^{\infty}{}_{n}\, b_n e^{i\mu_n t}$$

be respectively X-a.p., Y-a.p. functions. Moreover, let $\mathscr{S}_f$, $\mathscr{S}_g$ be the sets of all sequences regular with respect to $f(t)$, $g(t)$. In order that $\mathscr{S}_f\subseteq\mathscr{S}_g$, it is necessary and sufficient that each exponent μ_j be a linear combination, with integer coefficients, of a finite number of exponents λ_k.

The condition is sufficient. Let $\{s_n\}\in\mathscr{S}_f$ and, by assumption,

$$\mu_j = \sum_{1}^{N_j}{}_{k}\, c_{jk}\lambda_k \qquad (c_{jk} \text{ integers}).$$

Then

$$\mathscr{M}\{f(t+s_n)e^{-i\lambda_k t}\} = \lim_{T\to\infty}\frac{1}{2T}\int_{-T}^{T} f(t)e^{-i\lambda_k(t-s_n)}dt$$

$$= e^{i\lambda_k s_n}a_k \qquad (a_k\neq 0)$$

and, being uniformly

$$\lim_{n\to\infty} f(t+s_n) = f_s(t),$$

we have

$$\lim_{n\to\infty} e^{i\lambda_k s_n}a_k = \mathscr{M}(f_s(t)e^{-i\lambda_k t}).$$

Since $a_k \neq 0$, it follows that, $\forall k$,

$$\lim_{n\to\infty} e^{i\lambda_k s_n} = e^{i\alpha_k}$$

exists. Therefore,

$$\lambda_k s_n = \alpha_k + 2\pi p_{kn} + \rho_{kn},$$

with p_{kn} integer, $\lim_{n\to\infty} \rho_{kn} = 0$. Hence

$$e^{i\mu_j s_n} = \exp\left(i \sum_{1}^{N_j}{}_k c_{jk}\lambda_k s_n\right) \to \exp\left(i \sum_{1}^{N_j}{}_k c_{jk}\alpha_k\right).$$

Therefore, $\{s_n\} \in S_{Q_m}$, $\forall m$, where $Q_m(t)$ is the approximation polynomial (defined in theorem V) of $g(t)$, such that

$$\sup_J \|g(t) - Q_m(t)\| \leq \frac{1}{m} \qquad (m = 1, 2, \cdots).$$

From this we immediately deduce that $\{s_n\} \in \mathscr{S}_g$.

The condition is necessary. Let $\mathscr{S}_f \subseteq \mathscr{S}_g$ and suppose, by absurd assumption, that there exists an exponent μ_j which is not a linear combination with integer coefficients of a finite number of exponents λ_k.

There are two possibilities, (γ_1) and (γ_2):

(γ_1) There exists an integer $l_j > 1$ such that

$$l_j\mu_j = \sum_{1}^{N}{}_k d_{jk}\lambda_k, \tag{5.17}$$

with l_j and d_{jk} relatively prime integers.

(γ_2) Such an integer does not exist, that is, no multiple of μ_j can be obtained as a finite linear combination, with integer coefficients, of the λ_k's. In this case any finite linear combination, $= 0$, of the λ_k's and of μ_j has the form

$$\sum_{1}^{n}{}_k d_{jk}\lambda_k = 0 \qquad (\Leftrightarrow l_j = 0). \tag{5.18}$$

Let us consider case (γ_1). Let $l'_j \geq 2$ be the smallest possible positive integer for which (5.17) holds. We shall prove that each l_j is divisible by l'_j. Assume in fact this is not true; then $l_j = m_j l'_j + h_j$ with $1 \leq h_j < l'_j$. Hence, $h_j\mu_j$ can be expressed as a finite combination, with integer coefficients, of the λ_k's, which is absurd.

Setting

$$l'_j\mu_j = \sum_{1}^{N'}{}_k d'_{jk}\lambda_k, \tag{5.19}$$

let us consider, for $N \geq N'$, $\delta > 0$ arbitrary, the system of $N+1$ inequalities, (mod 2π), in the unknown τ,

$$(5.20)\qquad \begin{aligned} &|\lambda_k \tau| \leq \delta \qquad (k = 1, \cdots, N) \\ &\left|\mu_j \tau - \frac{2\pi}{l'_j}\right| \leq \delta. \end{aligned}$$

This system has, by Kronecker's theorem, a solution τ when we set

$$\lambda_{N+1} = \mu_j; \qquad \theta_1 = \cdots = \theta_N = 0, \qquad \theta_{N+1} = \frac{2\pi}{l'_j}.$$

In this case, in fact, each relation of the form

$$\sum_{1}^{N}{}_k \, d_{jk}\lambda_k = l_j \mu_j \qquad (d_{jk},\, l_j \text{ integers})$$

implies

$$l_j \frac{2\pi}{l'_j} = m_j 2\pi = 0 \pmod{2\pi}.$$

Let us now consider the function $g(t) \sim \sum_{(1)}^{(\infty)}{}_n \, b_n e^{i\mu_n t}$. We have

$$\begin{aligned} \sup_J \|g(t+\tau) - g(t)\| &= \sup_J \|(g(t+\tau) - g(t)) e^{-i\mu_j t}\| \\ &\geq \|\mathscr{M}(g(t+\tau) - g(t)) e^{-i\mu_j t}\| \\ &= \|b_j\| \, |e^{i\mu_j \tau} - 1|. \end{aligned}$$

By the last part of (5.20), we obtain

$$\frac{2\pi}{l'_j} - \delta \leq \mu_j \tau \leq \frac{2\pi}{l'_j} + \delta \pmod{2\pi}$$

and, consequently, taking $\delta \leq \pi / l'_j$, we have

$$\frac{\pi}{l'_j} \leq \mu_j \tau \leq \frac{3\pi}{l'_j} \pmod{2\pi}.$$

Hence,

$$|e^{i\mu_j \tau} - 1| \geq |e^{i(\pi/l'_j)} - 1| = \rho_j > 0,$$

that is,

$$(5.21)\qquad \sup_J \|g(t+\tau) - g(t)\| \geq \rho_j \|b_j\| > 0.$$

It may be noted that the right-hand side of (5.21) does not depend on N. Let us now consider the translation function:

$$v_f(\tau) = \sup_J \|f(t+\tau) - f(t)\|,$$

and define a polynomial $P_m(t)$ such that

$$\sup_J \|f(t) - P_m(t)\| \leq \frac{1}{m}.$$

Then

$$v_f(\tau) \leq \frac{2}{m} + \sup_J \|P_m(t+\tau) - P_m(t)\|, \tag{5.22}$$

$$P_m(t) = \sum_{1}^{n_m}{}_k\, a_{mk} e^{i\lambda_k t} \qquad (\|a_{mk}\| \leq M = \sup_J \|f(t)\|). \tag{5.23}$$

Let us now take $\varepsilon > 0$ arbitrarily and (assuming $m \geq 4/\varepsilon$, $\delta \leq \varepsilon/2n_m M$) let τ be a solution of system (5.20) with $N = n_m$. We obtain, by (5.23),

$$\|P_m(t+\tau) - P_m(t)\| \leq M \sum_{1}^{n_m}{}_k\, |e^{i\lambda_k \tau} - 1| \leq M n_m \delta \leq \frac{\varepsilon}{2}$$

and, by (5.22),

$$v_f(\tau) \leq \frac{\varepsilon}{2} + \frac{\varepsilon}{2} = \varepsilon.$$

For the same τ it is, on the other hand, by (5.21),

$$v_g(\tau) \geq \rho_j \|b_j\|.$$

It follows (considering the comparison function $\omega_{f,g}(\varepsilon)$) that

$$\omega_{f,g}(\varepsilon) \geq \rho_j \|b_j\|$$

and, consequently,

$$\omega_{f,g}(0+) \geq \rho_j \|b_j\|.$$

From theorem X of Chapter I it follows that $\mathscr{S}_f \nsubseteq \mathscr{S}_g$ and the theorem is proved.

Let us now consider, for the case (γ_2), the system

$$\begin{rcases} |\lambda_k \tau| \leq \delta \\ |\mu_j \tau - \pi| \leq \delta \end{rcases} \pmod{2\pi} \qquad (k = 1, \cdots, N).$$

This admits a solution, since any linear combination, $= 0$, with integer coefficients, of the λ_k's and of μ_j has the form (5.18); consequently $l_j = 0$ ($\Rightarrow l_j \pi = 0 \pmod{2\pi}$).

Therefore, we obtain

$$\pi - \delta \leq \mu_j \tau \leq \pi + \delta \pmod{2\pi}$$

and, for $\delta \leq \pi/2$,

$$\frac{\pi}{2} \leq \mu_j \tau \leq \frac{3\pi}{2} \pmod{2\pi},$$

$$|e^{i\mu_j \tau} - 1| \geq |e^{i(\pi/2)} - 1| = \sqrt{2}$$

We can then proceed as for the case (γ_1).

XI. *Let $\{\lambda_n\}$ be the sequence of characteristic exponents of $f(t)$. If we assume $\varepsilon>0$ arbitrarily, there exists a positive integer N_ε and a number $\delta_\varepsilon>0$ such that each solution τ of the system of inequalities*

$$|e^{i\lambda_n\tau}-1| \leq \delta_\varepsilon \qquad (n = 1,\cdots, N_\varepsilon) \tag{5.24}$$

is an ε-a.p. for $f(t)$.

It may be noted that system (5.24) has solutions for any $\delta_\varepsilon>0$ and any N_ε. Furthermore, by (5.23),

$$\begin{aligned}\|f(t+\tau)-f(t)\| &\leq \|f(t+\tau)-P_m(t+\tau)\|+\|P_m(t+\tau)-P_m(t)\| \\ &\quad +\|P_m(t)-f(t)\| \\ &\leq \frac{2}{m}+\sum_{1}^{n_m}{}_k \|a_{mk}\|\ |e^{i\lambda_k\tau}-1|.\end{aligned}$$

Taking m_ε such that $2/m_\varepsilon\leq\varepsilon/2$ and setting $N_\varepsilon=n_{m_\varepsilon}$, let τ be a solution of the system

$$|e^{i\lambda_k\tau}-1| \leq \varepsilon\left(\sum_{1}^{N_\varepsilon}{}_k \|a_{m_\varepsilon k}\|\right)^{-1} = \delta_\varepsilon.$$

Then, for the same τ,

$$\sup_J \|f(t+\tau)-f(t)\| \leq \varepsilon$$

and the theorem is proved.

CHAPTER 3

WEAKLY ALMOST-PERIODIC FUNCTIONS

1. DEFINITION OF W.A.P. FUNCTION

Let X be a Banach space and X^* the dual space. Let x denote the elements of X, x^* the elements of X^* (continuous linear functionals on X), $\|x\|$ and $\|x^*\|$ the respective norms.

The function, with complex values, defined on X, obtained by applying the functional x^* to the vector x, will be indicated by the symbol

$$\langle x^*,x\rangle.$$

Owing to the linearity with respect to both x and x^*, we obtain

$$\left\langle \sum_{1}^{n}{}_k\, \alpha_k x_k^*, \sum_{1}^{n}{}_j\, \beta_j x_j \right\rangle = \sum_{1}^{n}{}_k \sum_{1}^{n}{}_j\, \alpha_k \beta_j \langle x_k^*, x_j\rangle,$$

for any choice of the complex numbers α_k, β_j.

Furthermore, let

$$x = f(t) \qquad (t \in J)$$

be a function with values in X.

We shall say that $f(t)$ is weakly almost-periodic (w.a.p.) if the function $\langle x^*, f(t)\rangle$ is a.p., $\forall x^* \in X^*$.

The definition given is, obviously, related to that of a.p. function as the definition of weakly continuous function (i.e. such that $\langle x^*, f(t)\rangle$ is continuous, $\forall x^* \in X^*$) is to that of continuous function.

In the present chapter we shall develop the theory of w.a.p. functions (Amerio [5], [8]; the definition given here of weak almost-periodicity and theorem X can also be found in Kopec [1]. A different definition of weak almost-periodicity is due to Eberlein [1], [2]: the w.a.p. functions in the sense of Eberlein possess notable properties, particularly in relation to ergodic theorems).

Let $\{x_n\}$ be any sequence. We shall say that $\{x_n\}$ is *scalarly convergent* if the numerical sequence $\{\langle x^*,x_n\rangle\}$ converges $\forall x^*$. If $\{x_n\}$ is scalarly convergent and if, in addition, there *exists* $x \in X$ such that, $\forall x^* \in X^*$,

$$\lim_{n\to\infty} \langle x^*,x_n\rangle = \langle x^*,x\rangle,$$

then the sequence $\{x_n\}$ will be said to be *weakly convergent*; the value x will be called the *weak limit* of $\{x_n\}$, using the notation

$$\lim_{n\to\infty}{}^{*} x_n = x \quad \text{or} \quad x_n \xrightarrow{*} x.$$

It is well known, by Hahn-Banach's theorem, that, if the weak limit exists, it is unique.

If the space X is such that every scalarly convergent sequence is also weakly convergent, then X is said to be *weakly sequentially complete*, or *semicomplete*.

Reflexive spaces are semicomplete; hence the Hilbert spaces and the spaces l^p and L^p, with $1<p<+\infty$, are semicomplete. The space C^0, of numerical continuous functions on the interval $0 \vdash\!\dashv 1$, is *not* semicomplete.

By the notation

$$\lim_{n\to\infty} x_n = x \quad \text{or} \quad x_n \to x$$

we shall, as usual, indicate that the sequence $\{x_n\}$ converges (in the strong sense) to the value x, i.e., that $\|x_n - x\| \to 0$.

2. FIRST PROPERTIES OF W.A.P. FUNCTIONS

I. *$f(t)$ a.p. $\Rightarrow$ $f(t)$ w.a.p.*

In fact

$$|\langle x^*, f(t+\tau)-f(t)\rangle| \le \|x^*\| \, \|f(t+\tau)-f(t)\|.$$

II. *$f(t)$ w.a.p. $\Rightarrow$ $\mathscr{R}_{f(t)}$ bounded and separable.*

In fact, $\forall x^*$, the function $\langle x^*, f(t)\rangle$ is a.p. and consequently bounded; it follows that $f(t)$ is bounded ($\sup_J \|f(t)\| < +\infty$). The range $\mathscr{R}_{f(t)}$ is separable because $f(t)$, w.a.p., is weakly continuous.

III. *$f_n(t)$ w.a.p. $(n=1, 2, \cdots)$ and $\lim^*_{n\to\infty} f_n(t) = f(t)$ uniformly $\Rightarrow$ $f(t)$ w.a.p.*

Since

$$\lim_{n\to\infty} \langle x^*, f_n(t)\rangle = \langle x^*, f(t)\rangle \qquad (\forall x^* \in X^*)$$

uniformly, and since $\langle x^*, f_n(t)\rangle$ is a.p., $\forall x^*$, it follows that $\langle x^*, f(t)\rangle$ is a.p., $\forall x^*$; this means that $f(t)$ is w.a.p.

IV. *Let $f(t)$ be w.a.p. and $s=\{s_n\}$ a sequence such that, $\forall t \in J$,*

$$\lim_{n\to\infty}{}^{*} f(t+s_n) = g(t). \tag{2.1}$$

The convergence is then uniform, so that $g(t)$ is w.a.p. Furthermore, if we call $\Omega_{f(t)}$ the convex extension of $\mathscr{R}_{f(t)}$, we have

$$\overline{\Omega}_{f(t)} = \overline{\Omega}_{g(t)} \tag{2.2}$$

and

$$\sup_J \|f(t)\| = \sup_J \|g(t)\|. \tag{2.3}$$

As $\langle x^*, f(t)\rangle$ is a.p., it follows, by theorem XI of Chapter 1, that the sequence $\{\langle x^*, f(t+s_n)\rangle\}$ converges uniformly to $\langle x^*, g(t)\rangle$; hence, by theorem III, $g(t)$ is w.a.p.

Let us now recall that $\overline{\Omega}_{f(t)}$ is the closure of the set $\Omega_{f(t)}$ constituted by the points

$$z = \sum_{1}^{p}{}_j \, \rho_j f(t_j),$$

p being an arbitrary positive integer, $t_1, \cdots, t_p$ arbitrary points $\in J$ and $\rho_1, \cdots, \rho_p$ constants satisfying the conditions

$$\rho_k \geq 0, \qquad \sum_{1}^{p}{}_k \, \rho_k = 1.$$

Consider an arbitrary point of $\Omega_{g(t)}$:

$$y = \sum_{1}^{p}{}_k \, \rho_k g(t_k).$$

Then, by (2.1),

$$y = \lim_{n\to\infty}{}^* \sum_{1}^{p}{}_k \, \rho_k f(t_k+s_n) = \lim_{n\to\infty}{}^* z_n,$$

where

$$z_n = \sum_{1}^{p}{}_k \, \rho_k f(t_k+s_n) \in \Omega_{f(t)}.$$

By a theorem of Mazur, $\overline{\Omega}_{f(t)}$, closed and convex, is also weakly closed. Therefore, $y \in \overline{\Omega}_{f(t)}$; hence

$$\overline{\Omega}_{g(t)} \subseteq \overline{\Omega}_{f(t)}.$$

We observe now that, if we fix x^* arbitrarily, we have

$$\lim_{n\to\infty} \langle x^*, f(t+s_n)\rangle = \langle x^*, g(t)\rangle$$

uniformly. Therefore, for any $\varepsilon > 0$, there exists n_ε (depending on x^*) such that, for $n \geq n_\varepsilon$,

$$\sup_J |\langle x^*, f(t+s_n)\rangle - \langle x^*, g(t)\rangle| \leq \varepsilon.$$

Hence,

$$\sup_J |\langle x^*, f(t)\rangle - \langle x^*, g(t-s_n)\rangle| \leq \varepsilon.$$

We obtain, therefore, uniformly,

$$\lim_{n\to\infty} \langle x^*, g(t-s_n)\rangle = \langle x^*, f(t)\rangle$$

that is, x^* being arbitrary,

$$\lim_{n\to\infty}{}^* g(t-s_n) = f(t) \tag{2.4}$$

uniformly.

It follows that

$$\overline{\Omega}_{f(t)} \subseteq \overline{\Omega}_{g(t)}$$

and, consequently, (2.2).

From (2.1) and (2.4) we have, finally,

$$\|g(t)\| \leq \min \lim_{n\to\infty} \|f(t+s_n)\| \leq \sup_J \|f(t)\|,$$

$$\|f(t)\| \leq \sup_J \|g(t)\|,$$

which proves also (2.3).

Observation I. Let Z denote the linear space constituted by the functions $f(t)$, with values in X, weakly continuous and bounded; every element $\tilde{f} \in Z$ will therefore be defined by such a function:

$$\tilde{f} = \{f(t);\, t \in J\}.$$

Observe now that Z is a linear topological space if we give the following definition of neighborhood (suggested by that corresponding to the weak topology on X).

Let us choose n arbitrary numbers $\varepsilon_k > 0$ $(k = 1, \cdots, n)$ and let $x_1^*, \cdots, x_n^*$ be n elements, also arbitrary, of X^*.

We shall call "*neighborhood*" $\mathcal{U}_{\tilde{f}}(x_1^*, \cdots, x_n^*; \varepsilon_1, \cdots, \varepsilon_n)$ of $\tilde{f}$ the set of the elements $\tilde{g} = \{g(t);\, t \in J\}$ of Z such that

$$\sup_J |\langle x_k^*, g(t)-f(t)\rangle| < \varepsilon_k \qquad (k = 1, \cdots, n).$$

It can then be easily seen that the definition of weak almost-periodicity given at § 1 is equivalent to the following.

Let $f(t) \in Z$ and set $\tilde{f}(\tau) = \{f(t+\tau);\, t \in J\}$, $\forall \tau \in J$. We shall say that $f(t)$ is w.a.p. if, for any arbitrary neighborhood $\mathcal{U}_{\tilde{f}}$ of $\tilde{f}$, there exists, correspondingly, a relatively dense set $\{\tau\}_{\mathcal{U}_{\tilde{f}}}$ such that

$$\tilde{f}(\tau) \in \mathcal{U}_{\tilde{f}}, \qquad \forall \tau \in \{\tau\}_{\mathcal{U}_{\tilde{f}}}.$$

3. HARMONIC ANALYSIS OF W.A.P. FUNCTIONS: APPROXIMATION THEOREM, UNIQUENESS THEOREM, AND EXTENSION OF BOCHNER'S CRITERION

We shall assume in this section that the space X is *semicomplete*. The following theorems then hold.

V. *If $f(t)$ is w.a.p. there exists, $\forall \lambda \in J$, the mean value*

$$\lim_{T\to\infty}{}^* \frac{1}{2T}\int_{-T}^{T} f(t)e^{-i\lambda t}dt = \mathscr{M}(f(t)e^{-i\lambda t}) = a(\lambda). \tag{3.1}$$

Furthermore,

$$\lim_{T\to\infty}{}^* \frac{1}{2T}\int_{-T+s}^{T+s} f(t)e^{-i\lambda t}dt = a(\lambda), \tag{3.2}$$

uniformly with respect to $s \in J$.

Observe, first of all, that the function $f(t)e^{-i\lambda t}$ is Riemann-integrable on every interval $(-T+s)\vdash\!\dashv(T+s)$, being weakly continuous.

The function $\langle x^*, f(t)\rangle$ is a.p., $\forall$ fixed x^*. The mean value

$$\lim_{T\to\infty} \frac{1}{2T}\int_{-T}^{T} \langle x^*, f(t)\rangle e^{-i\lambda t}dt = \lim_{T\to\infty} \left\langle x^*, \frac{1}{2T}\int_{-T}^{T} f(t)e^{-i\lambda t}dt\right\rangle \tag{3.3}$$

therefore exists. Being x^* arbitrary and X semicomplete, from (3.3) follows the existence of the weak limit

$$a(\lambda) = \lim_{T\to\infty}{}^* \frac{1}{2T}\int_{-T}^{T} f(t)e^{-i\lambda t}dt = \mathscr{M}(f(t)e^{-i\lambda t}).$$

Relation (3.1) is therefore proved. Furthermore, $\forall x^*$,

$$\lim_{T\to\infty} \frac{1}{2T}\int_{-T+s}^{T+s} \langle x^*, f(t)\rangle e^{-i\lambda t}dt = \langle x^*, a(\lambda)\rangle$$

uniformly with respect to s. Also (3.2) is then proved.

We shall call $a(\lambda)$ *the Bohr transform of the w.a.p. function $f(t)$*. For the sake of simplicity, we have written $a(\lambda)$ in place of $a(\lambda; f(t))$.

VI. *If $f(t)$ is w.a.p., then $a(\lambda)=0$ with the exception of, at most, a sequence $\{\lambda_n\}$.*

By theorem II, $\mathscr{R}_{f(t)} \subset X_0$, separable subspace of X; it follows that also $a(\lambda) \in X_0$. As X_0 is separable, there exists, as is known, a *determining* sequence, $\{x_r^*\}$, with respect to X_0; consequently,

$$\|a(\lambda)\| = \sup_r |\langle x_r^*, a(\lambda)\rangle|. \tag{3.4}$$

From the almost-periodicity of $\langle x_r^*, f(t)\rangle$ it follows that $\langle x_r^*, a(\lambda)\rangle = 0$ with the exception of, at most, a sequence $\{\lambda_{rk}\}$. Therefore, by (3.4), $a(\lambda) = 0$, with the exception, at most, of the sequence $\{\lambda_n\} = \bigcup_{r,k} \lambda_{rk}$.

Setting

$$a_n = a(\lambda_n), \tag{3.5}$$

let us associate to the function $f(t)$ the *Fourier expansion*:

$$f(t) \sim \sum_{1}^{\infty}{}_n\, a_n e^{i\lambda_n t}. \tag{3.6}$$

The constants λ_n are called the *characteristic exponents* of $f(t)$.

Let now $B = \{\beta_n\}$ be a basis for the sequence $\{\lambda_n\}$ and let us extend to the series (3.6) *Bochner's summation procedure*, generalizing in this way the *approximation theorem* to w.a.p. functions.

For this, consider the same polynomials $Q_m(f(t))$ defined by (3.5), (3.6), and (3.17) of Chapter 2.

We have

$$Q_m(f(t)) = \sum_{1}^{r_m}{}_k\, \mu_{mk} a_k e^{i\lambda_k t}, \tag{3.7}$$

where the convergence factors μ_{mk} depend *only* on the sequence of characteristic exponents $\{\lambda_n\}$ and satisfy the inequalities

$$0 < \mu_{mk} \leq 1.$$

VII. *The limit*

$$\lim_{n \to \infty}{}^* Q_m(f(t)) = f(t) \tag{3.8}$$

holds uniformly on J.

Observe that B is a basis for the sequence of characteristic exponents corresponding to the function $\langle x^*, f(t)\rangle$, for any choice of x^*. From (3.3) it follows, in fact, that

$$a(\lambda; \langle x^*, f(t)\rangle) = \mathscr{M}(\langle x^*, f(t)e^{-i\lambda t}\rangle) = \langle x^*, a(\lambda)\rangle$$

and, consequently, $a(\lambda; \langle x^*, f(t)\rangle) = 0$ for $\lambda \notin \{\lambda_n\}$.

Furthermore,

$$\langle x^*, Q_m(f(t))\rangle = \sum_{1}^{r_m}{}_k\, \mu_{mk} \langle x^*, a(\lambda_k)\rangle e^{i\lambda_k t},$$

that is, $\langle x^*, Q_m(f(t))\rangle$ is the mth Bochner polynomial constructed from the basis B and corresponding to the function $\langle x^*, f(t)\rangle$; this is also true if, for some k, $\langle x^*, a(\lambda_k)\rangle = 0$.

It follows that

$$\lim_{m \to \infty} \langle x^*, Q_m(f(t))\rangle = \langle x^*, f(t)\rangle \tag{3.9}$$

uniformly and (3.8) is proved.

VIII. $a(\lambda) \equiv 0 \Rightarrow f(t) \equiv 0$ *(uniqueness theorem).*

In fact, if $a(\lambda) = 0$, $\forall \lambda$, then, by (3.7), $Q_m(f(t)) = 0$, $\forall m$. From (3.8) it follows that $f(t) = 0$.

Finally, let us extend Bochner's criterion.

IX. *Let $f(t)$ be weakly continuous. Then $f(t)$ is w.a.p. if, and only if, from every sequence $\{l_n\}$ it is possible to select a subsequence $\{s_n\}$ such that the sequence $\{f(t+s_n)\}$ is uniformly weakly convergent.*

The sufficiency of this condition is obvious.

Let us prove that the condition is necessary. For this purpose, we select, from $\{l_n\}$, a subsequence $\{s_n\}$ such that, $\forall k$,

$$\lim_{n\to\infty} e^{i\lambda_k s_n} = e^{i\alpha_k}.$$

It follows that, $\forall m$,

$$\lim_{n\to\infty} Q_m(f(t+s_n)) = \sum_{1}^{r_m}{}_k \mu_{mk} a_k e^{i\alpha_k} e^{i\lambda_k t} \tag{3.10}$$

uniformly.

Furthermore, $\forall x^*$,

$$\begin{aligned} |\langle x^*, f(t+s_p)\rangle - \langle x^*, f(t+s_n)\rangle| \le{} & |\langle x^*, f(t+s_p)\rangle - \langle x^*, Q_m(f(t+s_p))\rangle| \\ & + |\langle x^*, Q_m(f(t+s_p))\rangle - \langle x^*, Q_m(f(t+s_n))\rangle| \\ & + |\langle x^*, Q_m(f(t+s_n))\rangle - \langle x^*, f(t+s_n)\rangle| \end{aligned}$$

and the thesis follows from (3.8) and (3.10).

Clearly, by theorem III, *the limit function*

$$f_s(t) = \lim_{n\to\infty}{}^* f(t+s_n)$$

is, like $f(t)$, w.a.p.

Observation II. Let Z be the topological vector space defined in observation I in § 2. Bochner's criterion can then be expressed in the following way.

The function $f(t)$ is w.a.p. if, and only if, the range $\mathscr{R}_{\tilde{f}(s)}$ of the function $\tilde{f}(s) = \{f(t+s);\, t \in J\}$ is, in Z, sequentially relatively compact.

4. CRITERIA FOR $f(t)$, W.A.P., TO BE A.P.

It is interesting to note that the hypothesis that must be added to weak almost-periodicity in order to obtain strong almost-periodicity is one of *compactness*. The following theorem, in fact, holds.

X. $f(t)$ *a.p.* $\Leftrightarrow f(t)$ *w.a.p. and* $\mathscr{R}_{f(t)}$ *r.c.*

The condition is obviously necessary.

Let us prove that it is sufficient. We shall show first that $f(t)$ (which is w.a.p. and has r.c. range) is continuous. Suppose, in fact, that t_0 is a point of discontinuity. There exist then a number $\sigma>0$ and two sequences $\{h_{n1}\}$, $\{h_{n2}\}$ such that

$$\lim_{n\to\infty} h_{n1} = 0, \qquad \lim_{n\to\infty} h_{n2} = 0,$$

$$\|f(t_0+h_{n1})-f(t_0+h_{n2})\| \geq \sigma. \tag{4.1}$$

As the range $\mathscr{R}_{f(t)}$ is r.c., we may assume (selecting eventually from $\{h_{n1}\}$, $\{h_{n2}\}$ two subsequences which will again be called $\{h_{n1}\}$, $\{h_{n2}\}$) that

$$\lim_{n\to\infty} f(t_0+h_{n1}) = a_1, \qquad \lim_{n\to\infty} f(t_0+h_{n2}) = a_2 \tag{4.2}$$

and, by (4.1), we shall have

$$\|a_1-a_2\| \geq \sigma. \tag{4.3}$$

By Hahn-Banach's theorem, there exists then a functional $y^* \in X^*$ such that

$$\langle y^*,a_1\rangle \neq \langle y^*,a_2\rangle. \tag{4.4}$$

Now $f(t)$, being w.a.p., is weakly continuous. By the continuity of $\langle y^*,f(t)\rangle$, we have then

$$\langle y^*,f(t_0)\rangle = \lim_{n\to\infty} \langle y^*,f(t_0+h_{n1})\rangle = \lim_{n\to\infty} \langle y^*,f(t_0+h_{n2})\rangle.$$

Hence, by (4.2),

$$\langle y^*,f(t_0)\rangle = \langle y^*,a_1\rangle = \langle y^*,a_2\rangle,$$

which contradicts (4.4).

The continuity of $f(t)$ is therefore proved. We now prove that $f(t)$ is a.p., and, to do this, we shall utilize Bochner's criterion.

Let $s=\{s_n\}$ be an arbitrary sequence. If $\{\eta_k\}$ is the sequence of rational numbers, we can assume (being $\mathscr{R}_{f(t)}$ r.c.) that the sequences $\{f(\eta_k+s_n)\}$ converge, $\forall k$; hence,

$$\lim_{n\to\infty} f(\eta_k+s_n) = g_k. \tag{4.5}$$

Let us show that the convergence of the sequence $\{f(t+s_n)\}$ is uniform; the thesis will then follow from Bochner's criterion.

Assume that the convergence is not uniform. There exist then a number $\sigma>0$ and three sequences

$$\{\xi_n\}, \qquad \{s_{n1}\} \subset \{s_n\}, \qquad \{s_{n2}\} \subset \{s_n\}, \tag{4.6}$$

such that

$$\|f(\xi_n+s_{n1})-f(\xi_n+s_{n2})\| \geq \sigma. \tag{4.7}$$

Moreover, by the relative compactness, we can assume that

$$(4.8)\qquad \lim_{n\to\infty} f(\xi_n+s_{n1}) = b_1, \qquad \lim_{n\to\infty} f(\xi_n+s_{n2}) = b_2,$$

where

$$(4.9)\qquad \|b_1-b_2\| \geq \sigma.$$

By Hahn-Banach's theorem, there exists a functional $z^* \in X^*$ such that

$$(4.10)\qquad \langle z^*,b_1\rangle \neq \langle z^*,b_2\rangle.$$

The function $\varphi(t)=\langle z^*,f(t)\rangle$ is a.p.; moreover, $\forall k$,

$$\varphi(\eta_k+s_n) = \langle z^*, f(\eta_k+s_n)\rangle \to \langle z^*,g_k\rangle.$$

From theorem XI of Chapter 1 it follows that

$$(4.11)\qquad \lim_{(m,n)\to\infty} \sup_J |\varphi(t+s_n)-\varphi(t+s_m)| = 0.$$

We can also assume, by Bochner's criterion, that (uniformly)

$$(4.12)\qquad \lim_{n\to\infty} \varphi(t+\xi_n+s_{ni}) = \psi_i(t) \qquad (i = 1, 2).$$

Let us prove that

$$(4.13)\qquad \psi_1(t) = \psi_2(t).$$

In fact, by (4.6) and (4.11),

$$\begin{aligned}
&|\varphi(t+\xi_n+s_{n1})-\varphi(t+\xi_n+s_{n2})| \\
&\qquad \leq |\varphi(t+\xi_n+s_{n1})-\varphi(t+\xi_n+s_n)|+|\varphi(t+\xi_n+s_n)-\varphi(t+\xi_n+s_{n2})| \\
&\qquad \leq \sup_J |\varphi(t+s_n)-\varphi(t+s_{n1})|+\sup_J |\varphi(t+s_n)-\varphi(t+s_{n2})| \to 0.
\end{aligned}$$

If $t=0$, it follows, from (4.8), (4.12), and (4.13), that

$$\begin{aligned}
\langle z^*,b_1\rangle &= \lim_{n\to\infty} \varphi(\xi_n+s_{n1}) = \psi_1(0) = \psi_2(0) \\
&= \lim_{n\to\infty} \varphi(\xi_n+s_{n2}) = \langle z^*,b_2\rangle,
\end{aligned}$$

which contradicts (4.10).

We now give two other criteria, which hold in spaces of rather particular nature, but are very important in theory and applications (for instance, *Hilbert spaces and the spaces* L^p, l^p, *with* $1<p<+\infty$).

Let us first prove the following *lemma*.

Assume that $f(t)$ *is w.a.p. and that, for a given sequence* $\{s_n\}$,

$$(4.14)\qquad \lim_{n\to\infty}{}^{*} f(t+s_n) = g(t)$$

uniformly. Then, if the norms $\|f(t)\|$, $\|g(t)\|$ *are a.p.*,

$$(4.15)\qquad \lim_{n\to\infty} \|f(t+s_n')\| = \|g(t)\|$$

uniformly, $\{s_n'\}$ *being a suitable subsequence of* $\{s_n\}$.

Since $\|f(t)\|$ is a.p., there exists a subsequence $s' \subseteq s$ such that

$$\lim_{n\to\infty} \|f(t+s'_n)\| = \varphi(t) \tag{4.16}$$

uniformly; $\varphi(t)$ is, moreover, a.p.

From (4.14) and (4.16) it follows that

$$\|g(t)\| \leq \lim_{n\to\infty} \|f(t+s'_n)\| = \varphi(t). \tag{4.17}$$

Moreover, by (4.14),

$$\lim_{n\to\infty}{}^{*} g(t-s_n) = f(t) \tag{4.18}$$

uniformly.

Since $\|g(t)\|$ is a.p., we have, uniformly,

$$\lim_{n\to\infty} \|g(t-s''_n)\| = \psi(t), \tag{4.19}$$

where s'' is a suitable subsequence of s'. Hence, by (4.18),

$$\|f(t)\| \leq \psi(t). \tag{4.20}$$

From (4.20) it follows that

$$\|f(t+s''_n)\| \leq \psi(t+s''_n)$$

and, consequently, by (4.16) and (4.19),

$$\varphi(t) = \lim_{n\to\infty} \|f(t+s''_n)\| \leq \lim_{n\to\infty} \psi(t+s''_n) = \|g(t)\|. \tag{4.21}$$

Relation (4.15) then follows from (4.16), (4.17), and (4.21).

The uniformity of the convergence is a consequence of theorem XI of Chapter 1, $\|f(t)\|$ being a.p.

Let us now prove the following criterion.

XI. *Assume that:*

(a) *X is semicomplete,*

(b) $x_n \overset{*}{\to} x$ *and* $\|x_n\| \to \|x\| \Rightarrow x_n \to x$.

Let $f(t)$ be w.a.p. and denote by $\mathscr{S}_f$ the set of all sequences $s = \{s_n\}$ regular with respect to $f(t)$ (i.e. such that $\lim^*_{n\to\infty} f(t+s_n) = f_s(t)$ uniformly). Assume, finally, that $\|f_s(t)\|$ is a.p., $\forall s \in \mathscr{S}_f$. Then $f(t)$ is a.p.

To prove the thesis it will be sufficient to show that the range $\mathscr{R}_{f(t)}$ is r.c.

If not, there would exist a number $\rho > 0$ and a sequence $l = \{l_n\}$ such that

$$\|f(l_j) - f(l_k)\| \geq \rho, \qquad (j \neq k). \tag{4.22}$$

By Bochner's criterion, we can select from l a subsequence $s = \{s_n\} \in \mathscr{S}_f$; it will therefore be

$$\lim_{n\to\infty}{}^{*} f(t+s_n) = f_s(t) \tag{4.23}$$

uniformly, with $f_s(t)$ w.a.p.

Since $\|f(t)\|$ is a.p., it is possible to select from s a subsequence $s' = \{s'_n\}$ such that the sequence $\{\|f(t+s'_n)\|\}$ converges uniformly. Hence, by the preceding lemma,

$$\lim_{n\to\infty} \|f(t+s'_n)\| = \|f_s(t)\|. \tag{4.24}$$

From (4.23) and (4.24), it follows, by property (b), that

$$\lim_{n\to\infty} \|f(t+s'_n) - f_s(t)\| = 0, \qquad \forall t \in J. \tag{4.25}$$

Relation (4.25), written for $t=0$, contradicts (4.22) and the theorem is proved.

The last criterion, which we shall now prove, holds for *uniformly convex spaces* (or *Clarkson* spaces). These spaces are defined in the following way.

X is called a *uniformly convex space* if, in the interval $0 < \sigma \le 2$, a function $\omega(\sigma)$, with $0 < \omega(\sigma) \le 1$, is defined such that, $\forall x_1$ and x_2, the inequalities

$$\|x_1\|, \|x_2\| \le 1, \qquad \|x_1 - x_2\| \ge \sigma, \tag{4.26}$$

imply

$$\left\|\frac{x_1+x_2}{2}\right\| \le 1 - \omega(\sigma). \tag{4.27}$$

Hence, $\forall x_1$, x_2 satisfying the condition

$$\|x_1 - x_2\| \ge \sigma \max\{\|x_1\|, \|x_2\|\}, \tag{4.28}$$

we obtain

$$\left\|\frac{x_1+x_2}{2}\right\| \le (1-\omega(\sigma)) \max\{\|x_1\|, \|x_2\|\}. \tag{4.29}$$

Uniformly convex spaces satisfy condition (b); moreover they are reflexive (and consequently semicomplete).

The spaces l^p, L^p with $1 < p < +\infty$, and the Hilbert spaces are uniformly convex.

Let us verify that a *Hilbert space is uniformly convex.*

It follows, in fact, from the parallelogram theorem, that

$$\|x_1 - x_2\|^2 + \|x_1 + x_2\|^2 = 2\{\|x_1\|^2 + \|x_2\|^2\}$$

and therefore, if (4.26) hold,

$$\|x_1 + x_2\|^2 \le 4 - \sigma^2,$$

which coincides with (4.27), provided we set

$$1 - \omega(\sigma) = \sqrt{1 - \frac{\sigma^2}{4}}. \tag{4.30}$$

We now prove the following criterion.

XII. *Let X be uniformly convex and $f(t)$ w.a.p. Assume that, for every sequence $s=\{s_n\}$ such that*

$$\|f(s_j)-f(s_k)\| \geq \rho > 0 \qquad (j \neq k) \tag{4.31}$$

(ρ depending on s), we have also, $\forall t \in J$,

$$\max_{(j,k)\to\infty} \lim \|f(t+s_j)-f(t+s_k)\| \geq \sigma_\rho > 0, \tag{4.32}$$

σ_ρ being independent of t. Then $f(t)$ is a.p.

It will be sufficient to prove that the range $\mathscr{R}_{f(t)}$ is r.c.

If not, there would exist a number $\rho>0$ and a sequence s such that (4.31) would hold.

We observe that $f(t)$ is w.a.p. and X semicomplete. By Bochner's criterion, we can then assume that

$$\lim_{n\to\infty}{}^{*} f(t+s_n) = f_s(t) \tag{4.33}$$

uniformly; moreover $f_s(t)$ is, like $f(t)$, w.a.p.

From (2.3) it follows that

$$\sup_J \|f(t)\| = \sup_J \|f_s(t)\| = M < +\infty. \tag{4.34}$$

For every fixed $t \in J$, there exist, by (4.32), two subsequences

$$s' = \{s'_n\} \subset \{s_n\}, \qquad s'' = \{s''_n\} \subset \{s_n\}$$

(which depend on t) such that

$$\|f(t+s'_n)-f(t+s''_n)\| \geq \frac{\sigma_\rho}{2}.$$

Hence, by (4.34),

$$\|f(t+s'_n)-f(t+s''_n)\| \geq \frac{\sigma_\rho}{2M} \max\{\|f(t+s'_n)\|, \|f(t+s''_n)\|\}. \tag{4.35}$$

Since X is uniformly convex, it follows, by (4.29), that

$$\begin{aligned} \left\|\frac{f(t+s'_n)+f(t+s''_n)}{2}\right\| &\leq \left(1-\omega\left(\frac{\sigma_\rho}{2M}\right)\right) \max\{\|f(t+s'_n)\|, \|f(t+s''_n)\|\} \\ &\leq \left(1-\omega\left(\frac{\sigma_\rho}{2M}\right)\right)M. \end{aligned} \tag{4.36}$$

Therefore, by (4.33),

$$\|f_s(t)\| \leq \left(1-\omega\left(\frac{\sigma_\rho}{2M}\right)\right)M,$$

contrary to (4.34).

Observation III. It may be noted that the thesis holds, even if (4.32) is verified $\forall$ fixed $t \geq t_0$ ($t \leq t_0$).

Observe, in fact, that, *if $f(t)$ is w.a.p., setting*

$$M^+ = \sup_{t \geq t_0} \|f(t)\|, \qquad M^- = \sup_{t \leq t_0} \|f(t)\|, \qquad M = \sup_J \|f(t)\|$$

we have $M^+ = M^- = M$.

Suppose, for instance, that $M^+ < M$; there exists then a point $\bar{t} < t_0$ such that $\|f(\bar{t})\| > M^+$ and, by Hahn-Banach's theorem, a functional $\bar{x}^* \in X^*$ such that

$$|\langle \bar{x}^*, f(\bar{t})\rangle| = \|f(\bar{t})\| > M^+, \qquad \|\bar{x}^*\| = 1. \tag{4.37}$$

Hence, by (4.37),

$$\begin{aligned} K = \sup_J |\langle \bar{x}^*, f(t)\rangle| \geq |\langle \bar{x}^*, f(\bar{t})\rangle| > M^+ = \sup_{t \geq t_0} \|f(t)\| \\ = \|\bar{x}^*\| \sup_{t \geq t_0} \|f(t)\| \geq \sup_{t \geq t_0} |\langle \bar{x}^*, f(t)\rangle| = k \geq 0. \end{aligned} \tag{4.38}$$

Relation (4.38) is absurd. Setting in fact $\delta = K - k > 0$ and taking $t_\delta \in J$ such that $|\langle \bar{x}^*, f(t_\delta)\rangle| \geq k + \frac{2}{3}\delta$, let us consider the set, $\{\tau\}$, of the $\delta/3$-a.p. of $\langle \bar{x}^*, f(t)\rangle$. It will be, $\forall \tau$, $|\langle \bar{x}^*, f(t_\delta + \tau)\rangle| \geq k + \delta/3$, which is absurd, since, for a suitable $\bar{\tau}$, $t_\delta + \bar{\tau} \geq t_0$.

It must, therefore, be $M^+ = M$ (and, analogously, $M^- = M$).

The proof of theorem XII can then be repeated, without any modification, assuming that (4.32) holds $\forall t \geq t_0$ (or $t \leq t_0$).

Observation IV. *Let us assume that X is a separable Hilbert space and let $\{z_n\}$ be a complete orthonormal sequence.* Then, if $f(t)$ takes its values in X, we have, $\forall t \in J$,

$$f(t) = \sum_{1}^{\infty}{}_n \varphi_n(t) z_n \tag{4.39}$$

where $\varphi_n(t) = (f(t), z_n)$ is the scalar product of $f(t)$ by z_n.

By (4.39), we have

$$\|f(t)\|^2 = \sum_{1}^{\infty}{}_n |\varphi_n(t)|^2. \tag{4.40}$$

Let us prove the following statements.

XIII. $f(t)$ *w.a.p.* $\Leftrightarrow \varphi_n(t)$ *a.p.*, $\sum_{(1)}^{(\infty)}{}_n |\varphi_n(t)|^2 \leq M^2 < +\infty$.

XIV. $f(t)$ *a.p.* $\Leftrightarrow \varphi_n(t)$ *a.p.*, $\sum_{(1)}^{(\infty)}{}_n |\varphi_n(t)|^2$ *uniformly convergent.*

The necessity of the condition of theorem XIII is obvious.

For its sufficiency, observe that, if we choose $y \in X$ arbitrarily, we have

$$y = \sum_{1}^{\infty}{}_n \eta_n z_n, \qquad \left(\eta_n = (y, z_n), \sum_{1}^{\infty}{}_n |\eta_n|^2 = \|y\|^2\right).$$

Consider the scalar product $(f(t),y)$; to prove the thesis it is sufficient to show that it is a.p. Now we have

$$(f(t),y) = \sum_{1}^{\infty}{}_n \varphi_n(t)\bar{\eta}_n \tag{4.41}$$

and the series on the right-hand side (constituted by a.p. functions) converges uniformly, since it is, by Schwarz's inequality,

$$\left|\sum_{p}^{q}{}_n \bar{\eta}_n\varphi_n(t)\right| \leq \left\{\sum_{p}^{q}{}_n |\eta_n|^2\right\}^{1/2}\left\{\sum_{p}^{q}{}_n |\varphi_n(t)|^2\right\}^{1/2} \leq M\left\{\sum_{p}^{q}{}_n |\eta_n|^2\right\}^{1/2}.$$

To prove that the condition expressed by theorem XIV is necessary, we observe that, $f(t)$ being a.p., the range $\mathscr{R}_{f(t)}$ is r.c. This implies that, $\forall \varepsilon > 0$, there exists a finite number of points $f(t_1), \cdots, f(t_\nu)$ such that

$$\mathscr{R}_{f(t)} \subset \bigcup_{j}^{1\cdots\nu} (f(t_j),\varepsilon).$$

Let us now take an index m_ε such that

$$\|f_{m_\varepsilon}(t_j)\| = \left\{\sum_{m_\varepsilon}^{\infty}{}_n |\varphi_n(t_j)|^2\right\}^{1/2} \leq \varepsilon \qquad (j = 1, \cdots, \nu).$$

Chosen arbitrarily $t' \in J$, we have, for a certain $t_{j'}$, $\|f(t')-f(t_{j'})\| < \varepsilon$; therefore

$$\begin{aligned}\left\{\sum_{m_\varepsilon}^{\infty}{}_n |\varphi_n(t')|^2\right\}^{1/2} &= \|f_{m_\varepsilon}(t')\| \leq \|f_{m_\varepsilon}(t_{j'})\| + \|f_{m_\varepsilon}(t')-f_{m_\varepsilon}(t_{j'})\| \\ &\leq \|f_{m_\varepsilon}(t_{j'})\| + \|f(t')-f(t_{j'})\| < 2\varepsilon.\end{aligned}$$

As t' is arbitrary, the thesis is proved.

It is obvious that the condition is sufficient since the uniform convergence of the series $\sum_{(1)}^{(\infty)}{}_n |\varphi_n(t)|^2$ is equivalent to the uniform convergence of the series (of a.p. functions) on the right-hand side of (4.39).

CHAPTER 4

THE INTEGRATION OF ALMOST-PERIODIC FUNCTIONS

1. INTRODUCTION AND STATEMENTS

Let X be a Banach space and $x = f(t)$ a continuous function from J to X. We shall set

$$F(t) = \int_0^t f(\eta)d\eta. \tag{1.1}$$

In Euclidean spaces Bohl-Bohr's theorem holds: *X Euclidean, $f(t)$ a.p., $F(t)$ bounded $\Rightarrow$ $F(t)$ a.p.*

The problem of extending this result to Banach spaces, which we shall now treat, serves as a model for the study of more general cases, which refer to the integration of abstract a.p. differential equations (in particular partial differential equations) considered in Part II.

We shall now state the main results that will be proved in the following sections.

First of all, the following theorem holds.

I (*Bochner* [1]). *X Banach space, $f(t)$ a.p., $\mathscr{R}_{F(t)}$ r.c. $\Rightarrow$ $F(t)$ a.p.*

As can be seen, the hypothesis of boundedness in the theorem of Bohl-Bohr is substituted by the much stricter hypothesis of compactness. This hypothesis cannot, however, as we shall see in an example (Amerio [9]), be substituted in the *general* case by one of boundedness.

Consider, in fact, the space $X = l^\infty$ of bounded sequences of complex numbers: $x = \{\xi_n\}$, with $\|x\| = \sup_n |\xi_n|$. The function, from J to X,

$$f(t) = \left\{\frac{1}{n}\cos\frac{t}{n}\right\} \tag{1.2}$$

is a.p. In fact, if we set

$$f_n(t) = \left\{\cos t, \cdots, \frac{1}{n}\cos\frac{t}{n}, 0, 0, \cdots\right\},$$

$f_n(t)$ is periodic, with period $2\pi(n!)$, and we have

$$\|f(t) - f_n(t)\| \leq \frac{1}{n+1}.$$

Moreover,

$$F(t) = \left\{\sin \frac{t}{n}\right\}. \tag{1.3}$$

Hence, $\|F(t)\| \leq 1$, but $F(t)$ is *not* a.p. Since, in fact,

$$\left|\sin \frac{t+\tau}{n} - \sin \frac{t}{n}\right| \leq \|F(t+\tau) - F(t)\|,$$

if $F(t)$ were a.p., the functions $\sin(t/n)$ would be equally a.p.; they would admit, in other words, $\forall \varepsilon > 0$, a common r.d. set $\{\tau\}_\varepsilon$ of a.p. This cannot occur if $\varepsilon < 2$. In fact,

$$\begin{aligned} \sup_J \left|\sin \frac{t+\tau}{n} - \sin \frac{t}{n}\right| &= \sup_J 2\left|\cos \frac{2t+\tau}{2n} \sin \frac{\tau}{2n}\right| \\ &= 2\left|\sin \frac{\tau}{2n}\right| \end{aligned}$$

and the condition

$$2\left|\sin \frac{\tau}{2n}\right| \leq \varepsilon$$

is equivalent to the condition

$$m\pi - \delta_\varepsilon \leq \frac{\tau}{2n} \leq m\pi + \delta_\varepsilon,$$

with $m = 0, \pm 1, \pm 2, \cdots$, $\sin \delta_\varepsilon = \varepsilon/2$, $0 < \delta_\varepsilon < \pi/2$.

Thus, if $\{\tau\}_{\varepsilon,n}$ is the set of the ε-a.p. of $\sin(t/n)$, then

$$\{\tau\}_{\varepsilon,n} = \bigcup_m 2n(m\pi - \delta_\varepsilon) \vdash\!\dashv 2n(m\pi + \delta_\varepsilon),$$

that is, $\{\tau\}_{\varepsilon,n}$ is constituted by a sequence of equal intervals, of length $4n\delta_\varepsilon$, with centers at the points $2nm\pi$. Moreover, the distance between two consecutive intervals is

$$2n\{(m+1)\pi - \delta_\varepsilon\} - 2n\{m\pi + \delta_\varepsilon\} = 2n(\pi - 2\delta_\varepsilon),$$

which $\to +\infty$ when $n \to +\infty$. Hence,

$$\{\tau\}_\varepsilon = \bigcap_n \{\tau\}_{\varepsilon,n} = -2\delta_\varepsilon \vdash\!\dashv 2\delta_\varepsilon,$$

that is, $\{\tau\}_\varepsilon$ is not r.d. and $F(t)$ cannot be a.p.

A similar example can be given for $X = C^0$, space of continuous numerical functions on the interval $0 \vdash\!\dashv 1$.

The problem now arises of determining whether there are some *particular* Banach spaces to which the theorem of Bohl-Bohr can be extended word by word. This actually can be done for notable spaces—for example, for

Hilbert spaces and, more generally, for uniformly convex spaces (Amerio [4], [9]).

The following statement, in fact, holds.

II (*Amerio* [9]). *X uniformly convex,* $f(t)$ *a.p.,* $F(t)$ *bounded* $\Rightarrow$ $F(t)$ *a.p.*

This property, which refers to the solutions of the very simple abstract equation $x'(t)=f(t)$, gives origin, in a natural way, to a classification of Banach spaces, with regard to the relation between *boundedness* and *almost-periodicity* of the solutions of differential a.p. equations; this relationships exists, as we shall see later, for notable equations, related to the mechanics of continuous systems and to theoretical physics.

The question as to whether theorem II can be extended to *reflexive* spaces is open. It must, however, be noted that the statement is true for some *not* reflexive spaces, for example, in l^1 (Ricci and Rizzonelli [1]). More generally, consider the space $X=l^p\{X_n\}$, with $1 \leq p < +\infty$, defined in the following way.

Let $\{X_n\}$ be a sequence of Banach spaces, x_n the elements of X_n, $\|x_n\|_{X_n}$ their norms. Having fixed p, with $1 \leq p < +\infty$, each element $x \in X$ is constituted by a *sequence*

$$x = \{x_n\},$$

with $x_n \in X_n$, assuming that $\sum_{(1)}^{(\infty)}{}_n \|x_n\|_{X_n}^p < +\infty$.

The norm in X is defined by

$$\|x\| = \left\{\sum_{1}^{\infty}{}_n \|x_n\|_{X_n}^p\right\}^{1/p}.$$

If $X_n=\mathscr{C}$ (complex field), then $X=l^p$.

Let us observe that $x=f(t)$, with values in X and continuous, means

$$f(t) = \{f_n(t)\}, \tag{1.4}$$

with $f_n(t)$ continuous from J to X_n. Furthermore,

$$F(t) = \int_0^t f(\eta)d\eta = \left\{\int_0^t f_n(\eta)d\eta\right\} = \{F_n(t)\}. \tag{1.5}$$

The following theorem now holds.

III (*Amerio* [11]). *If, in each space* X_n, *the following statement holds:*

$$f_n(t) \text{ a.p.}, F_n(t) \text{ bounded} \Rightarrow F_n(t) \text{ a.p.},$$

then the same statement holds for the space $X=l^p\{X_n\}$, *with* $1 \leq p < +\infty$:

$$f(t) \text{ a.p.}, F(t) \text{ bounded} \Rightarrow F(t) \text{ a.p.}$$

Theorem III holds, for example, if all the spaces X_n are uniformly convex.

For the proof, we shall use a theorem (Amerio [3]) that generalizes to a.p. functions Dini's well known theorem on monotonic sequences of continuous functions. The same statement makes it possible to give another proof for theorem II, when X is a Hilbert space (Amerio [4]; see the observation at the end of § 5).

Theorems II and III extend the theorem of Bohl-Bohr to Banach spaces of particular structure. Other theorems, which we shall now state, hold in any space and are obtained by assuming supplementary hypotheses on $f(t)$.

First of all, if, for $f(t)$, the Fourier expansion

$$f(t) \sim \sum_{1}^{\infty}{}_n\, a_n e^{i\lambda_n t} \tag{1.6}$$

holds, it is possible to extend (with the proof given by Levitan [2]) a theorem of Favard [1] on numerical a.p. functions (see Zaidman [3]).

IV (*Favard* [1]). *X Banach space,* $f(t)$ *a.p.,* $|\lambda_n| \geq \alpha > 0 \Rightarrow F(t)$ *a.p.*

It is clear that, if the value $\lambda_1 = 0$ is among the characteristic exponents, then $F(t)$ cannot be a.p. In fact, in this case,

$$\lim_{T\to\infty} \frac{1}{T}\int_0^T f(t)dt = \lim_{T\to\infty} \frac{F(T)}{T} = a_1 \neq 0 \tag{1.7}$$

and $F(t)$ is not bounded (in other words, for the equation $x'(t) = f(t)$, we have the phenomenon of *resonance*). The assumption that $\lambda_n \neq 0$, $\forall n$, is not, however, sufficient to guarantee the almost-periodicity of $F(t)$, as can be seen from example (1.2), where $\lambda_n = 1/n$.

Another statement has been obtained by Levitan [2], starting from the observation that the integral $F(t)$ given by (1.3), even if *bounded*, does *not* have *mean value*; in other words, there does not exist a value b such that

$$\lim_{T\to\infty} \frac{1}{T}\int_0^T F(t)dt = b. \tag{1.8}$$

We have, in fact,

$$\frac{1}{T}\int_0^T F(t)dt = \left\{\frac{n}{T}\left(1-\cos\frac{T}{n}\right)\right\}.$$

Setting

$$b = \{\beta_n\},$$

we obtain

$$\beta_n = \lim_{T\to\infty} \frac{n}{T}\left(1-\cos\frac{T}{n}\right) = 0 \Rightarrow b = 0,$$

whereas, setting $T = m$, with $m = 1, 2, \cdots$, we have

$$\left\|\frac{1}{m}\int_0^m F(t)dt\right\| = \sup_n \frac{n}{m}\left(1-\cos\frac{m}{n}\right) \geq 1-\cos 1 > 0.$$

Another example (Levitan [2]), always regarding the space l^∞, shows that even the existence of the mean value of the integral $F(t)$ (in addition to the boundedness of $F(t)$) is not sufficient to guarantee almost-periodicity. This last can be proved under stricter conditions, that is, by assuming that, for $F(t)$, (2.2) of Chapter 2 holds.

The following theorem is thus obtained.

V (*Levitan* [2]). *X Banach space, $f(t)$ a.p., $F(t)$ bounded,*

$$\lim_{T\to\infty} \frac{1}{T}\int_{\sigma}^{\sigma+T} F(t)dt = b \tag{1.9}$$

uniformly with respect to $\sigma \Rightarrow F(t)$ a.p.

We note, finally, that the questions treated in the present chapter, together with others, have been generalized by Günzler [3], [5], [6] to a.p. functions defined on a semigroup.

2. PROOF OF THEOREM I

(*a*) We shall prove theorem I assuming, at first, that $X = J$ (in other words, proving the Bohl-Bohr theorem). Let $x = f(t)$ be a *real a.p.* function and assume $F(t)$ *bounded*. The proof we shall give that $F(t)$ is a.p. is due to Favard [1] and is based on Bochner's criterion.

By assumption,

$$-\infty < m = \inf_J F(t) \leq \sup_J F(t) = M < +\infty. \tag{2.1}$$

Let $\mathscr{S}_f$ be the family of sequences $s = \{s_n\}$ *regular* with respect to $f(t)$, that is, such that

$$\lim_{n\to\infty} f(t+s_n) = f_s(t), \tag{2.2}$$

uniformly. Then, by (1.1),

$$F(t+s_n) = F(s_n) + \int_0^t f(\eta+s_n)d\eta \tag{2.3}$$

and we can assume that

$$\lim_{n\to\infty} F(s_n) = C_s, \qquad \text{constant}. \tag{2.4}$$

From (2.2), (2.3), and (2.4), it follows that, $\forall t \in J$,

$$\lim_{n\to\infty} F(t+s_n) = C_s + \int_0^t f_s(\eta)d\eta = F_s(t), \tag{2.5}$$

so that $F_s(t)$ is an integral function of $f_s(t)$. Hence, by (2.1) and (2.5),

$$m \leq m_s = \inf_J F_s(t) \leq \sup_J F_s(t) = M_s \leq M. \tag{2.6}$$

Let us prove that

$$m_s = m, \qquad M_s = M. \tag{2.7}$$

Suppose, in fact, that

$$\sup_J F_s(t) = M_s < M. \tag{2.8}$$

From (2.2) it follows, by the uniform convergence, that

$$\lim_{n\to\infty} f_s(t-s_n) = f(t)$$

uniformly. We can then select from s a subsequence (again indicated by s) such that

$$\lim_{n\to\infty} F_s(t-s_n) = C+\int_0^t f(\eta)d\eta = C+F(t),$$

with C constant. Furthermore, by (2.6) and (2.8),

$$\inf_J (C+F(t)) = C+m \geq m_s \geq m,$$

$$\sup_J (C+F(t)) = C+M \leq M_s < M,$$

which is absurd.

We now prove that $F(t)$ is a.p. Let $s=\{s_n\}$ be an arbitrary sequence.

We shall assume (by Bochner's criterion) that $s\subset\mathscr{S}_f$ and that (2.2) therefore holds. Let us prove that the sequence $\{F(t+s_n)\}$ converges uniformly. If not, there would exist a constant $\rho>0$ and three sequences

$$\{\alpha_n\}, \qquad \{s_{n1}\}\subseteq\{s_n\}, \qquad \{s_{n2}\}\subseteq\{s_n\},$$

such that

$$|F(\alpha_n+s_{n2})-F(\alpha_n+s_{n1})| \geq \rho. \tag{2.9}$$

We may assume $\{\alpha_n+s_{n1}\}\in\mathscr{S}_f$ and $\{\alpha_n+s_{n2}\}\in\mathscr{S}_f$, that is,

$$\begin{aligned} &\lim_{n\to\infty} f(t+\alpha_n+s_{n1}) = f_1(t),\\ &\lim_{n\to\infty} f(t+\alpha_n+s_{n2}) = f_2(t), \end{aligned} \tag{2.10}$$

uniformly.

We shall further assume that, $\forall t\in J$,

$$\begin{aligned} &\lim_{n\to\infty} F(t+\alpha_n+s_{n1}) = F_1(t),\\ &\lim_{n\to\infty} F(t+\alpha_n+s_{n2}) = F_2(t), \end{aligned} \tag{2.11}$$

$F_1(t)$ and $F_2(t)$ being, respectively, integral functions of $f_1(t)$ and $f_2(t)$.

Let us prove that

$$(2.12) \qquad f_1(t) = f_2(t).$$

In fact, by (2.2),

$$\sup_J |f(t+\alpha_n+s_{n1})-f(t+\alpha_n+s_{n2})| = \sup_J |f(t+s_{n1})-f(t+s_{n2})| \to 0.$$

Hence,

$$F_2(t)-F_1(t) = C = \text{const}$$

and, by (2.9) and (2.11) (written for $t=0$),

$$(2.13) \qquad |C| \geq \rho > 0,$$

which contradicts the equalities

$$\sup_J F_2(t) = \sup_J F_1(t) = M.$$

The sequence $\{F(t+s_n)\}$ therefore converges uniformly and $F(t)$ is a.p.

(*b*) Let us prove the following *lemma*:

X Banach space, $f(t)$ *w.a.p.*, $F(t)=\int_0^t f(\eta)d\eta$ *bounded* $\Rightarrow$ $F(t)$ *w.a.p.*

Setting

$$(2.14) \qquad \sup_J \| F(t)\| = M < +\infty,$$

we obtain, in fact, $\forall x^* \in X^*$,

$$\left|\int_0^t \langle x^*,f(\eta)\rangle d\eta\right| = \left|\langle x^*, \int_0^t f(\eta)d\eta\rangle\right| = |\langle x^*,F(t)\rangle| \leq \|x^*\| M.$$

As $\langle x^*,f(t)\rangle$ is a.p., it follows then, by Bohl-Bohr's theorem, that $\langle x^*,F(t)\rangle$ is a.p., $\forall x^*$, which proves the lemma.

(*c*) The proof of theorem I is now straightforward. In fact, for what has been proved in (*b*), the function $F(t)$ is w.a.p. Assuming that the range $\mathscr{R}_{F(t)}$ is r.c., from theorem X of Chapter 3 it follows that $F(t)$ is a.p.

OBSERVATION I. Let us prove that, *if* $f(t)$, *a.p. from J to X*, *Banach space, has an a.p. integral function* $F(t)$, *then the corresponding Bohr transform satisfies the relations*

$$(2.15) \qquad a(0,f) = \mathscr{M}(f(t)) = 0, \qquad a(\lambda,f) = i\lambda a(\lambda,F) \quad \text{for} \quad \lambda \neq 0.$$

In fact,

$$a(\lambda,f) = \lim_{T\to\infty} \frac{1}{T} \int_{-(T/2)+s}^{(T/2)+s} e^{-i\lambda t} f(t)dt$$

uniformly with respect to s. Hence, setting $s=T/2$, we have

$$(2.16) \qquad a(\lambda,f) = \lim_{T\to\infty} \frac{1}{T} \int_0^T e^{-i\lambda t} f(t)dt.$$

Analogously,

$$a(\lambda,F) = \lim_{T\to\infty} \frac{1}{T}\int_0^T e^{-i\lambda T}F(t)dt. \tag{2.17}$$

From (2.16), with $\lambda=0$, and from (2.14), it follows that

$$a(0,f) = \lim_{T\to\infty} \frac{F(T)}{T} = 0.$$

Furthermore, for $\lambda\neq 0$, integrating by parts, we obtain

$$\begin{aligned} a(\lambda,f) &= \lim_{T\to\infty} \frac{1}{T}\int_0^T e^{-i\lambda t}f(t)dt \\ &= \lim_{T\to\infty} \frac{F(T)e^{-i\lambda T}}{T} + i\lambda \lim_{T\to\infty}\frac{1}{T}\int_0^T e^{-i\lambda t}F(t)dt \\ &= i\lambda a(\lambda,F), \end{aligned}$$

that is, the second of (2.15).

Hence, in the Fourier expansion of $f(t)$:

$$f(t) \sim \sum_{1}^{\infty}{}_n\, a_n e^{i\lambda_n t}, \tag{2.18}$$

it will necessarily be $\lambda_n\neq 0$; furthermore,

$$F(t) \sim C + \sum_{1}^{\infty}{}_n \frac{a_n}{i\lambda_n} e^{i\lambda_n t} \tag{2.19}$$

with C constant (with value in X).

Observe that expansion (2.19) is formally obtained by integrating term by term the series on the right-hand side of (2.18); moreover,

$$C = a(0,F) = \mathcal{M}(F(t)).$$

3. PROOF OF THEOREM II

It has been proved (in § 2, (*b*)) that $F(t)$ *bounded* $\Rightarrow$ $F(t)$ *w.a.p.* To prove theorem II it will therefore be sufficient to show that, if X is uniformly convex, *the range* $\mathscr{R}_{F(t)}$ *is r.c.* For this, we shall use theorem XII of Chapter 3.

Assume that there exists $\rho>0$ and a sequence $s=\{s_n\}$ such that

$$\|F(s_j)-F(s_k)\| \geq \rho > 0 \qquad (j\neq k). \tag{3.1}$$

We may then assume (by Bochner's criterion) that $s\in\mathscr{S}_f$, that is,

$$\lim_{(j,k)\to\infty} \|f(t+s_j)-f(t+s_k)\| = 0 \tag{3.2}$$

uniformly. Then, by (1.1) and (3.1), for fixed $t \in J$,

$$\|F(t+s_j)-F(t+s_k)\| = \left\|F(s_j)-F(s_k)+\int_0^t (f(\eta+s_j)-f(\eta+s_k))d\eta\right\| \geq \rho - \left\|\int_0^t (f(\eta+s_j)-f(\eta+s_k))d\eta\right\|. \tag{3.3}$$

From (3.2) and (3.3) it follows that

$$\max \lim_{(j,k)\to\infty} \|F(t+s_j)-F(t+s_k)\| \geq \rho$$

and the theorem is therefore proved.

4. MONOTONIC SEQUENCES OF A.P. FUNCTIONS

Let $\{\varphi_n(t)\}$ be a *monotonic and bounded* sequence of real functions, *continuous on the closed and bounded interval* $a \vdash\!\dashv b$. If, for instance, the sequence is decreasing, it will be

$$\varphi_1(t) \geq \varphi_2(t) \geq \cdots \geq \varphi_n(t) \geq \cdots \geq k > -\infty.$$

Dini's classical theorem states that, setting

$$\Phi(t) = \lim_{n\to\infty} \varphi_n(t),$$

if $\Phi(t)$ *is continuous on* $a \vdash\!\dashv b$, *then the sequence* $\{\varphi_n(t)\}$ *converges uniformly on the same interval.*

In this section we shall deal with the extension of Dini's theorem to sequences of a.p. functions (Amerio [3]).

Let $\{\varphi_n(t)\}$ *be a monotonic and bounded sequence of real a.p. functions; for instance*

$$\varphi_1(t) \geq \varphi_2(t) \geq \cdots \geq \varphi_n(t) \geq \cdots \geq k > -\infty. \tag{4.1}$$

Setting

$$\Phi(t) = \lim_{n\to\infty} \varphi_n(t), \tag{4.2}$$

one may ask if the assumption that the limit function $\Phi(t)$ is a.p. implies the uniform convergences (on J) of the sequence $\{\varphi_n(t)\}$.

This is *not* true, as can be seen through examples (see the observation at the end of this section). It is necessary, in order to prove that the convergence is uniform, to assume the almost-periodicity of a family $\{\Phi_s(t)\}$ of functions (which we shall define later) similarly to what was done in theorem XI of Chapter 3.

Let $s=\{s_m\}$ be a *regular* sequence with respect to *all* functions $\varphi_n(t)$, i.e., such that

(4.3) $$\lim_{m\to\infty} \varphi_n(t+s_m) = \varphi_{s,n}(t), \qquad \forall n,$$

uniformly.

Let $\mathscr{S}$ be the family of such sequences. If $l=\{l_m\}$ is an *arbitrary* sequence, it can be immediately shown, applying Bochner's criterion and Cantor's diagonal process, that *l contains a subsequence $s\in\mathscr{S}$.*

From (4.1) and (4.3), it also follows

(4.4) $$\varphi_{s,1}(t) \geq \varphi_{s,2}(t) \geq \cdots \geq \varphi_{s,n}(t) \geq \cdots \geq k > -\infty$$

and consequently there exists, $\forall t\in J$, the limit

(4.5) $$\Phi_s(t) = \lim_{n\to\infty} \varphi_{s,n}(t) \qquad (\Phi_0(t) = \Phi(t)).$$

The following theorem now holds.

VI. *If $\Phi_s(t)$ is a.p., $\forall s\in\mathscr{S}$, then the sequence $\{\varphi_n(t)\}$ converges uniformly on J.*

Proof. We can assume that $\Phi(t)=0$; otherwise, we could consider, in place of $\{\varphi_n(t)\}$, the sequence $\{\varphi_n(t)-\Phi(t)\}$.

Assume the convergence is not uniform. Setting

$$\rho_n = \sup_J \varphi_n(t)$$

we obtain then

$$\rho_n \geq \rho_{n+1}, \qquad \lim_{n\to\infty} \rho_n = \rho > 0.$$

Let us observe that, $\forall m$, there exists l_m such that

$$\varphi_m(l_m) \geq \frac{\rho}{2}.$$

Hence, by (4.1),

(4.6) $$\varphi_n(l_m) \geq \frac{\rho}{2}, \qquad \text{when } n \leq m.$$

We may now select from $l=\{l_m\}$ a subsequence $s=\{s_m\}\in\mathscr{S}$: it will be $s_m=l_{m+r_m}$, with $r_m\geq 0$, and therefore, by (4.6),

(4.7) $$\varphi_n(s_m) \geq \frac{\rho}{2}, \qquad \text{when } n \leq m.$$

It follows that

(4.8) $$\varphi_{s,n}(0) = \lim_{m\to\infty} \varphi_n(s_m) \geq \frac{\rho}{2}.$$

(4.9) $$\Phi_s(0) = \lim_{n\to\infty} \varphi_{n,s}(0) \geq \frac{\rho}{2}.$$

Observe that $\varphi_n(t) \geq 0 \Rightarrow \varphi_{s,n}(t) \geq 0 \Rightarrow \Phi_s(t) \geq 0$.

Let $s' = \{s'_m\}$ be a subsequence of s such that $\{-s'_m\}$ is regular with respect to $\Phi_s(t)$ (which, by assumption, is a.p.). Hence,

$$\begin{aligned}\lim_{n\to\infty} \Phi_s(t-s'_m) &= \Psi(t) \geq 0,\\ \lim_{n\to\infty} \Psi(t+s'_m) &= \Phi_s(t),\end{aligned} \tag{4.10}$$

uniformly. Since

$$\Phi_s(t-s'_m) \leq \varphi_{s,n}(t-s'_m),$$

it follows, when $m \to \infty$, by (4.3) and the first of (4.10), that

$$0 \leq \Psi(t) \leq \varphi_n(t).$$

Hence (letting $n \to \infty$) $\Psi(t) \equiv 0$, and, by the second of (4.10),

$$\Phi_s(t) \equiv 0,$$

which contradicts (4.9). The theorem is so proved.

This theorem has been generalized by Bochner [4] to *almost-automorphic* functions. For another proof, see Dolcher [1].

Observation II. The hypothesis that $\Phi_s(t)$ be a.p., $\forall s \in \mathscr{S}$, cannot be eliminated.

Suppose, in fact, that $\varphi_n(t)$ is an even and continuous function, periodic with period $2T_n = (4\pi)3^n$ (so that $T_{n+1} = 3T_n = T_n + 2T_n$), defined by setting

$$\varphi_n(t) = \begin{cases} 0 & \text{when} \quad 0 \leq t \leq T_n - \dfrac{\pi}{2} \\ \cos t & \text{when} \quad T_n - \dfrac{\pi}{2} < t \leq T_n. \end{cases} \tag{4.11}$$

By (4.11), $\varphi_n(t)$ vanishes on the whole of J with the exception of the intervals,

$$J_{n,r} = \left\{\left(T_n(1+2r) - \frac{\pi}{2}\right) \longmapsto \left(T_n(1+2r) + \frac{\pi}{2}\right)\right\} \quad (r = 0, +1, \ldots),$$

where it coincides with the function $\cos t$.

Therefore, $\forall t \in J$,

$$\Phi(t) = \lim_{n\to\infty} \varphi_n(t) = 0,$$

but, as $\varphi_n(T_n) = 1$, the convergence is *not* uniform on J.

Furthermore, $\forall$ integer r,

$$3(1+2r) = (1+2q_r),$$

with q_r integer. Hence,

$$J_{n+1,r} = \left\{\left(T_{n+1}(1+2r)-\frac{\pi}{2}\right)^{\longmapsto}\left(T_{n+1}(1+2r)+\frac{\pi}{2}\right)\right\}$$
$$= \left\{\left(T_n(1+2q_r)-\frac{\pi}{2}\right)^{\longmapsto}\left(T_n(1+2q_r)+\frac{\pi}{2}\right)\right\} = J_{n,q_r}$$

and therefore, on the whole of J,

$$\varphi_n(t) \geq \varphi_{n+1}(t) \geq 0.$$

From theorem VI it follows that, although $\Phi(t)$ is a.p. (as it vanishes identically), *not all* the $\Phi_s(t)$ can be a.p. This case appears, for example, if the sequence $\{s_m\}$ considered at (4.3) is defined by $s_m = T_m$. In fact, when $m \geq n$, then $\varphi_n(t+T_m) = \varphi_n(t+3^{m-n}T_n) = \varphi_n(t+(2q_{m,n}+1)T_n) = \varphi_n(t+T_n)$. Hence,

$$\varphi_{n,s}(t) = \varphi_n(t+T_n)$$

which implies, by (4.11), that

$$\Phi_s(t) = \begin{cases} \cos t & \text{for } |t| \leq \dfrac{\pi}{2} \\ 0 & \text{for } |t| > \dfrac{\pi}{2}. \end{cases}$$

5. PROOF OF THEOREM III

(*a*) By assumption, we have

$$f(t) = \{f_n(t)\}, \qquad f(t) \text{ a.p.},$$

and

$$\|f(t)\| = \left\{\sum_1^\infty {}_n \|f_n(t)\|_{X_n}^p\right\}^{1/p}, \qquad 1 \leq p < +\infty. \tag{5.1}$$

From the inequality

$$\|f_n(t+\tau)-f_n(t)\|_{X_n} \leq \|f(t+\tau)-f(t)\|,$$

it follows that the components $f_n(t)$ are (equally) a.p. Furthermore,

$$F(t) = \int_0^t f(\eta)d\eta = \{F_n(t)\}, \qquad \text{with} \quad F_n(t) = \int_0^t f_n(\eta)d\eta, \tag{5.2}$$

and also

(5.3) $$\|F(t)\| = \left\{\sum_1^\infty {}_n \|F_n(t)\|_{X_n}^p\right\}^{1/p} \leq M < +\infty,$$

so that

$$\|F_n(t)\|_{X_n} \leq M.$$

It follows, by assumption, that $F_n(t)$ is a.p., $\forall n$.

As the functions $\|F_n(t)\|_{X_n}^p$, $\|F(t)\|^p$ are continuous, the series (5.3) converges uniformly on every bounded interval (by Dini's theorem, recalled at § 4).

Let us now prove that the series (5.1) converges uniformly (on J). In fact, $f(t)$ being a.p., the range $\mathscr{R}_{f(t)}$ is r.c. If, therefore, we choose $\varepsilon>0$ arbitrarily, we can determine a finite number of values

$$f(t_1), \cdots, f(t_\nu)$$

such that, $\forall t \in J$,

(5.4) $$f(t) \in \bigcup_j^{1\ldots\nu} (f(t_j),\varepsilon).$$

If we set

$$g_k(t) = \{f_1(t), \cdots, f_k(t), 0, 0, \cdots\},$$

$g_k(t)$ takes its value in X and

(5.5) $$\left\{\sum_{k+1}^\infty {}_n \|f_n(t)\|_{X_n}^p\right\}^{1/p} = \|f(t) - g_k(t)\|.$$

It is then possible to choose k_ε in such a way that

(5.6) $$\|f(t_j) - g_{k_\varepsilon}(t_j)\| \leq \varepsilon \qquad \text{when} \quad j = 1, 2, \cdots, \nu.$$

For every $t' \in J$, there exists, by (5.4), an index j' such that

(5.7) $$\|f(t') - f(t_{j'})\| < \varepsilon.$$

Moreover (for any k, t_1, t_2), we have, obviously,

$$\|g_k(t_2) - g_k(t_1)\| \leq \|f(t_2) - f(t_1)\|.$$

From (5.6) and (5.7) it follows that

$$\begin{aligned}\|f(t') - g_{k_\varepsilon}(t')\| &\leq \|f(t') - f(t_{j'})\| + \|f(t_{j'}) - g_{k_\varepsilon}(t_{j'})\| \\ &\quad + \|g_{k_\varepsilon}(t_{j'}) - g_{k_\varepsilon}(t')\| \\ &\leq 2\|f(t') - f(t_{j'})\| + \|f(t_{j'}) - g_{k_\varepsilon}(t_{j'})\| < 3\varepsilon.\end{aligned}$$

As t' is arbitrary, the uniform convergence of the series (5.1) follows then from (5.5).

(*b*) Let us prove that the function $\|F(t)\|^p$ (which, by assumption, is bounded) is a.p.

First, we shall show that $\|F(t)\|^p$ is uniformly continuous (and even satisfies a Lipschitz condition). In fact, for $p=1$,

$$(5.8)\qquad |\,\|F(t+\tau)\|-\|F(t)\|\,| \le \|F(t+\tau)-F(t)\| = \left\|\int_t^{t+\tau} f(\eta)d\eta\right\| \le m|\tau|,$$

where

$$(5.9)\qquad m = \sup_J \|f(t)\|.$$

Furthermore, when $p>1$,

$$(5.10)\qquad |\,\|F(t+\tau)\|^p-\|F(t)\|^p| \le pM^{p-1}m|\tau|$$

and the uniform continuity is proved.

Now let $\zeta(t)\in C^1(J)$, satisfying the conditions

$$(5.11)\qquad \zeta(t)\begin{cases}\ge 0 & \text{when } |t|<1,\\ = 0 & \text{when } |t|\ge 1,\end{cases}\quad \int_J \zeta(t)dt = 1.$$

Then, for $k=1,2,\cdots$,

$$(5.12)\qquad k\int_J \zeta(kt)dt = \int_J \zeta(t)dt = 1.$$

Setting

$$(5.13)\qquad \omega_k(t) = k\int_J \zeta(k\tau)\|F(t-\tau)\|^p d\tau = k\int_{-1/k}^{1/k} \zeta(k\tau)\|F(t-\tau)\|^p d\tau,$$

we have

$$(5.14)\qquad \lim_{k\to\infty} \omega_k(t) = \|F(t)\|^p$$

uniformly.

In fact, by (5.10) and (5.12),

$$\begin{aligned}|\,\|F(t)\|^p-\omega_k(t)| &\le k\int_{-1/k}^{1/k} \zeta(k\tau)|\,\|F(t)\|^p-\|F(t-\tau)\|^p|d\tau\\ &\le kpM^{p-1}m\int_{-1/k}^{1/k} \zeta(k\tau)|\tau|d\tau \le \frac{pM^{p-1}m}{k}.\end{aligned}$$

If, therefore, we prove that $\omega_k(t)$ is a.p., $\forall k$, we may deduce that $\|F(t)\|^p$ is also a.p. Since, by (5.3), (5.11), and (5.13), $0\le\omega_k(t)\le M$, it will be sufficient to prove that the derivative $\omega_k'(t)$ is a.p., $\forall k$: then $\omega_k(t)$ will be a.p., by Bohl-Bohr's theorem.

By (5.12) and the uniform convergence of the series (5.3) on every bounded interval, it is

$$\omega_k(t) = \sum_1^{\infty}{}_n\, k \int_{-1/k}^{1/k} \zeta(k\tau)\|F_n(t-\tau)\|_{X_n}^p d\tau = \sum_1^{\infty}{}_n\, k \int_J \zeta(k\tau)\|F_n(t-\tau)\|_{X_n}^p d\tau$$

$$= \sum_1^{\infty}{}_n\, k \int_J \zeta(k(t-\tau))\|F_n(\tau)\|_{X_n}^p d\tau.$$

(5.15)

We now observe that, if $\tau \neq 0$, we obtain

$$\text{(5.16)} \quad \left|\frac{\|F_n(t+\tau)\|_{X_n} - \|F_n(t)\|_{X_n}}{\tau}\right| \leq \left\|\frac{F_n(t+\tau)-F_n(t)}{\tau}\right\|_{X_n}$$

$$= \left\|\frac{1}{\tau}\int_t^{t+\tau} f_n(\eta)d\eta\right\|_{X_n} \leq \sup_J \|f_n(t)\|_{X_n}.$$

Therefore $\|F_n(t)\|_{X_n}$ satisfies a Lipschitz condition and is differentiable almost-everywhere on J; precisely,

$$\text{(5.17)} \qquad \left|\frac{d}{dt}\|F_n(t)\|_{X_n}\right| \leq \|f_n(t)\|_{X_n}.$$

Hence, if $p>1$,

$$\text{(5.18)} \qquad \left|\frac{d}{dt}\|F_n(t)\|_{X_n}^p\right| \leq p\|F_n(t)\|_{X_n}^{p-1}\|f_n(t)\|_{X_n}.$$

It follows then, when $p>1$, that

$$\text{(5.19)} \quad \left|\frac{d}{dt} k\int_J \zeta(k(t-\tau))\|F_n(\tau)\|_{X_n}^p d\tau\right|$$

$$= \left|k^2\int_J \zeta'(k(t-\tau))\|F_n(\tau)\|_{X_n}^p d\tau\right| = \left|k\int_J \zeta(k\tau)\frac{d}{dt}\|F_n(t-\tau)\|_{X_n}^p d\tau\right|$$

$$\leq pk\int_{-1/k}^{1/k} \zeta(k\tau)\|F_n(t-\tau)\|_{X_n}^{p-1}\|f_n(t-\tau)\|_{X_n} d\tau,$$

whereas, if $p=1$,

$$\text{(5.20)} \quad \left|k^2\int_J \zeta'(k(t-\tau))\|F_n(\tau)\|_{X_n} d\tau\right| \leq k\int_{-1/k}^{1/k}\zeta(k\tau)\|f_n(t-\tau)\|_{X_n} d\tau.$$

From Hölder's inequality, it follows also, when $p>1$, that

$$\sum_r^s{}_n \|F_n(t)\|_{X_n}^{p-1}\|f_n(t)\|_{X_n} \leq \left\{\sum_r^s{}_n \|F_n(t)\|_{X_n}^p\right\}^{(p-1)/p}\left\{\sum_r^s{}_n \|f_n(t)\|_{X_n}^p\right\}^{1/p}$$

$$\leq M^{p-1}\left\{\sum_r^s{}_n \|f_n(t)\|_{X_n}^p\right\}^{1/p}$$

Hence, by the uniform convergence of the series (5.1), the series

$$\sum_1^\infty {}_n \|F_n(t)\|_{X_n}^{p-1} \|f_n(t)\|$$

also converges uniformly.

Consequently, the series

$$\sum_1^\infty {}_n k \int_{-1/k}^{1/k} \zeta(k\tau) \|F_n(t-\tau)\|_{X_n}^{p-1} \|f_n(t-\tau)\|_{X_n} d\tau \qquad \text{(when } p > 1\text{)},$$

$$\sum_1^\infty {}_n k \int_{-1/k}^{1/k} \zeta(k\tau) \|f_n(t-\tau)\|_{X_n} d\tau \qquad \text{(when } p = 1\text{)},$$

converge uniformly.

This implies the uniform convergence of the series

$$\sum_1^\infty {}_k \frac{d}{dt}\left(k \int_{-1/k}^{1/k} \zeta(k\tau) \|F_n(t-\tau)\|_{X_n}^p d\tau\right) \qquad (p \geq 1).$$

Hence, $\omega_k(t)$ is differentiable and

$$\omega_k'(t) = \sum_1^\infty {}_n k^2 \int_J \zeta'(k\tau) \|F_n(t-\tau)\|_{X_n}^p d\tau$$

$$= \sum_1^\infty {}_n k \int_J \zeta(k\tau) \frac{d}{dt} \|F_n(t-\tau)\|_{X_n}^p d\tau,$$

the series on the right-hand side being uniformly convergent.

Observe now that, $\|F_n(t)\|_{X_n}^p$ being a.p., also the function

$$\int_J \zeta'(k\tau) \|F_n(t-\tau)\|_{X_n}^p d\tau$$

is a.p. We have, in fact,

$$\left| \int_{-1/k}^{1/k} \zeta'(k\tau)(\|F_n(t+\delta-\tau)\|_{X_n}^p - \|F_n(t-\tau)\|_{X_n}^p) d\tau \right|$$

$$\leq (\sup_J | \|F_n(t+\delta)\|_{X_n}^p - \|F_n(t)\|_{X_n}^p |) \int_{-1/k}^{1/k} |\zeta'(k\tau)| d\tau.$$

The derivate $\omega_k'(t)$ is therefore a.p.

(*c*) It has been proved, in (*b*), that the series of a.p. nonnegative functions

(5.21) $$\sum_1^\infty {}_n \|F_n(t)\|_{X_n}^p \qquad (p \geq 1)$$

has a.p. sum $\|F(t)\|^p$.

Let us now show that the series (5.21) converges uniformly: for this we shall utilize theorem VI.

Let $s=\{s_k\}$ be a sequence *regular* with respect to all functions $\|F_n(t)\|_{X_n}$. Since $F_n(t)$ is a.p., we can assume that

$$\lim_{k\to\infty} F_n(t+s_k) = F_{s,n}(t) \qquad (\forall n), \tag{5.22}$$

uniformly. We can also assume that

$$\lim_{k\to\infty} f_n(t+s_k) = f_{s,n}(t) \qquad (\forall n), \tag{5.23}$$

uniformly. Moreover, $F_{s,n}(t)$ and $f_{s,n}(t)$ are a.p.

From (5.22) and (5.23) it follows, $\forall t \in J$, that

$$F_{s,n}(t) = \lim_{k\to\infty} \int_0^{t+s_k} f_n(\eta)d\eta = \lim_{k\to\infty} \left(F_n(s_k)+\int_0^t f_n(\eta+s_k)d\eta\right)$$

$$= F_{s,n}(0)+\int_0^t f_{s,n}(\eta)d\eta.$$

Moreover,

$$\left\{\sum_1^r{}_n \|F_{s,n}(t)\|_{X_n}^p\right\}^{1/p} \le \sup_J \left\{\sum_1^r{}_n \|F_n(t)\|_{X_n}^p\right\}^{1/p} \le M$$

and, consequently,

$$\left\{\sum_1^\infty{}_n \|F_{s,n}(t)\|_{X_n}^p\right\}^{1/p} \le M.$$

Hence, the function

$$F_s(t) = \{F_{s,n}(t)\} = \left\{F_{s,n}(0)+\int_0^t f_{s,n}(\eta)d\eta\right\}$$

takes its values in X (it even is $\|F_s(t)\| \le M$) and

$$F_s(t) = F_s(0)+\int_0^t f_s(\eta)d\eta.$$

Repeating the procedure followed in (*b*), it can be proved that the norm

$$\|F_s(t)\| = \left\{\sum_1^\infty{}_n \|F_{s,n}(t)\|_{X_n}^p\right\}^{1/p}$$

is a.p.

By theorem VI we conclude then that the series (5.3) converges uniformly.

Since, $\forall k$, the function

$$G_k(t) = \{F_1(t), \cdots, F_k(t), 0, 0, \cdots\}$$

is X-a.p., and as

$$\|F(t) - G_k(t)\| = \left\{\sum_{k+1}^{\infty}{}_n \|F_n(t)\|_{X_n}^p\right\}^{1/p},$$

it follows that $F(t)$ is X-a.p.

OBSERVATION III. *Let X be a Hilbert space and $f(t)$ a.p. from J to X.* As the range $\mathscr{R}_{f(t)}$ is separable, we can assume that X is separable. Let $\{u_n\}$ be an orthonormal sequence, complete on X; then, $\forall x \in X$,

$$x = \sum_1^{\infty}{}_n \xi_n u_n \qquad (\xi_n = (x, u_n)),$$

$$\|x\| = \left\{\sum_1^{\infty}{}_n |\xi_n|^2\right\}^{1/2}.$$

We may assume therefore that, if $\mathscr{C}$ is the complex field, $X = l^2\{\mathscr{C}\}$.

The proof of theorem II can therefore be deduced, if X is a Hilbert space, from that of theorem III. Moreover, in this case, the part of the proof given under (*a*) and (*b*) can be simplified (see Amerio [4]); we have, in fact,

$$f(t) = \sum_1^{\infty}{}_n \varphi_n(t) u_n \qquad (\varphi_n(t) = (f(t), u_n)),$$

(5.24)

$$F(t) = \sum_1^{\infty}{}_n \left(\int_0^t \varphi_n(\eta) d\eta\right) u_n,$$

with

(5.25)
$$\|F(t)\| = \left\{\sum_1^{\infty}{}_n \left|\int_0^t \varphi_n(\eta) d\eta\right|^2\right\}^{1/2} \leq M.$$

The functions $\varphi_n(t)$ are a.p. By (5.25) and by Bohl-Bohr's theorem, the integral functions

$$\int_0^t \varphi_n(\eta) d\eta$$

are also a.p. Hence, by the second of (5.24), in order to prove the almost-periodicity of $F(t)$ we have to show that the series of nonnegative a.p. functions

(5.26)
$$\sum_1^{\infty}{}_n \left|\int_0^t \varphi_n(\eta) d\eta\right|^2$$

converges uniformly.

In order to apply theorem VI, we prove first that $\|F(t)\|^2$, sum of series (5.26), is a.p. Observe, in fact, that

$$\frac{d}{dt}\|F(t)\|^2 = (F(t), f(t)) + (f(t), F(t)).$$

Hence, if we show that this derivative is a.p., the thesis will follow from Bohl-Bohr's theorem ($\|F(t)\|^2$ being bounded by (5.25)).

Let us recall that $F(t)$ *is w.a.p.* It will therefore be sufficient to prove that, *if* $F(t)$ *is w.a.p. and* $g(t)$ *a.p., the scalar product*

$$(F(t), g(t)) = \Psi(t)$$

is a.p.

Observe that

$$(5.27)\qquad |\Psi(t+\tau)-\Psi(t)| \leq |(F(t+\tau), g(t+\tau)-g(t))| + |(F(t+\tau)-F(t), g(t))|,$$

where, setting $M = \sup_J \|F(t)\|$,

$$(5.28)\qquad |F(t+\tau), g(t+\tau)-g(t))| \leq M \sup_J \|g(t+\tau)-g(t)\|.$$

The range $\mathscr{R}_{g(t)}$ being r.c., there exist, $\forall\varepsilon > 0$, ν values

$$g(t_1), \cdots, g(t_\nu)$$

such that, $\forall t \in J$,

$$(5.29)\qquad g(t) \in \bigcup_{k}^{1\cdots\nu} (g(t_k), \varepsilon).$$

Observe now that the $\nu+1$ functions

$$g(t),\ (F(t), g(t_1)), \cdots, (F(t), g(t_\nu))$$

are a.p.; they admit therefore a common set $\{\tau\}_\varepsilon$, r.d., of ε-a.p.

If $\tau \in \{\tau\}_\varepsilon$, then by (5.28), $\forall t \in J$,

$$(5.30)\qquad |(F(t+\tau), g(t+\tau)-g(t))| \leq M\varepsilon.$$

Having chosen $\bar{t} \in J$, let us take $\bar{k}$ in such a way that, by (5.29),

$$(5.31)\qquad \|g(\bar{t})-g(t_{\bar{k}})\| < \varepsilon.$$

Hence, if $\tau \in \{\tau\}_\varepsilon$,

$$|(F(\bar{t}+\tau)-F(\bar{t}), g(\bar{t}))|$$
$$\leq |(F(\bar{t}+\tau)-F(\bar{t}), g(t_{\bar{k}}))| + |(F(\bar{t}+\tau)-F(\bar{t}), g(\bar{t})-g(t_{\bar{k}}))| < \varepsilon + 2M\varepsilon.$$

From (5.27), (5.28), (5.30), and (5.31), we have, $\forall\tau \in \{\tau\}_\varepsilon$,

$$|\Psi(\bar{t}+\tau)-\Psi(\bar{t})| < (1+3M)\varepsilon$$

and, $\bar{t}$ being arbitrary, the thesis is proved.

It is obvious that, more generally, *if $F(t)$ is w.a.p. from J to X (Banach space) and if $g(t)$ is a.p. from J to the dual space X^*, the function*

$$\Psi(t) = \langle g(t), F(t) \rangle$$

is a.p.

6. PROOFS OF THEOREMS IV AND V

(*a*) The proofs we shall give, due to Levitan, are based on the Fourier transform. First, let us prove the following *lemma*.

Let $f(t) \sim \sum_{(1)}^{(\infty)}{}_n a_n e^{i\lambda_n t}$ a.p. from J to X, Banach space.

Moreover, let $\psi(s) \in L^1(J)$. Then the convolution

$$f * \psi(t) = \int_J f(t-\eta)\psi(\eta)d\eta = \int_J f(\eta)\psi(t-\eta)d\eta \tag{6.1}$$

is a.p. and has the Fourier expansion

$$f * \psi(t) \sim \sum_1^{\infty}{}_n a_n \varphi(\lambda_n) e^{i\lambda_n t}, \tag{6.2}$$

where $\varphi(\lambda)$ is the Fourier transform of $\psi(s)$:

$$\varphi(\lambda) = \int_J \psi(s) e^{-i\lambda s} ds.$$

Since, by (6.1),

$$\|f * \psi(t+\tau) - f * \psi(t)\| \leq \sup_J \|f(t+\tau) - f(t)\| \int_J |\psi(\eta)| d\eta,$$

from the almost-periodicity of $f(t)$ follows that of $f * \psi(t)$. Moreover, by theorem II of Chapter 2,

$$\begin{aligned}
\mathscr{M}(f * \psi(t) e^{-i\lambda t}) &= \lim_{T\to\infty} \frac{1}{T} \int_0^T e^{-i\lambda t}\, dt \int_J f(t-\eta)\psi(\eta) d\eta \\
&= \int_J \psi(\eta) e^{-i\lambda\eta} d\eta \lim_{T\to\infty} \frac{1}{T} \int_0^T f(t-\eta) e^{-i\lambda(t-\eta)} dt \\
&= \int_J \psi(\eta) e^{-i\lambda\eta} d\eta \lim_{T\to\infty} \frac{1}{T} \int_{-\eta}^{T-\eta} f(t) e^{-i\lambda t} dt \\
&= \varphi(\lambda) \mathscr{M}(f(t) e^{-i\lambda t}),
\end{aligned}$$

which proves (6.2).

(*b*) *Let us prove theorem IV*. Let $\varphi(\lambda) \in C^2(J)$, with

$$\varphi(\lambda) = \frac{1}{i\lambda}, \qquad \text{when } |\lambda| \geq \alpha > 0. \tag{6.3}$$

By (6.3) there exists, when $s \neq 0$, the inverse Fourier transform of $\varphi(\lambda)$:

$$\psi(s) = \frac{1}{2\pi} \int_J \varphi(\lambda) e^{i\lambda s} d\lambda;$$

integrating twice by parts, we have

$$\psi(s) = -\frac{1}{2\pi} \int_J \varphi'(\lambda) \frac{e^{i\lambda s}}{is} d\lambda = \frac{1}{2\pi} \int_J \varphi''(\lambda) \frac{e^{i\lambda s}}{(is)^2} d\lambda.$$

Hence

$$(6.4) \qquad |\psi(s)| \leq \frac{1}{2\pi s^2} \int_J |\varphi''(\lambda)| d\lambda.$$

Since

$$\psi(s) = \frac{1}{2\pi} \int_{-\alpha}^{\alpha} \varphi(\lambda) e^{i\lambda s} d\lambda + \frac{1}{\pi} \int_{\alpha}^{+\infty} \frac{\sin \lambda s}{\lambda} d\lambda,$$

then, when $s > 0$,

$$\psi(s) = \frac{1}{2\pi} \int_{-\alpha}^{\alpha} \varphi(\lambda) e^{i\lambda s} d\lambda + \frac{1}{\pi} \int_{\alpha s}^{+\infty} \frac{\sin \eta}{\eta} d\eta$$

and, when $s < 0$,

$$\psi(s) = \frac{1}{2\pi} \int_{-\alpha}^{\alpha} \varphi(\lambda) e^{i\lambda s} d\lambda - \frac{1}{\pi} \int_{-\alpha s}^{+\infty} \frac{\sin \eta}{\eta} d\eta.$$

The function $\psi(s)$ has, therefore, at $s = 0$, a discontinuity of the first kind, with jump $\psi(0^+) - \psi(0^-) = 1$; by (6.4) $\psi(s) \in L^1(J)$. Setting

$$G(t) = f * \psi(t),$$

$G(t)$ is then a.p. and has the Fourier expansion

$$G(t) \sim \sum_{1}^{\infty}{}_n\, a_n \varphi(\lambda_n) e^{i\lambda_n t}.$$

As $\mathscr{M}(f(t)e^{-i\lambda t}) = 0$ when $|\lambda| \leq \alpha$, it follows, by (6.3), that

$$G(t) \sim \sum_{1}^{\infty}{}_n\, a_n \frac{e^{i\lambda_n t}}{i\lambda_n},$$

that is, the expansion of $G(t)$ can be obtained by integrating term by term that of $f(t)$.

If we show that

$$(6.5) \qquad G'(t) = f(t),$$

the theorem will be proved (since $F(t) = G(t) - G(0)$).

Let $\{f_n(t)\}$ be a sequence of trigonometric polynomials

$$f_n(t) = \sum_{1}^{r_n}{}_k\, a_{nk} e^{i\lambda_k t}$$

uniformly convergent to $f(t)$.

Setting

$$G_n(t) = \int_J f_n(t-s)\psi(s)ds,$$

we have

$$G_n(t) = \sum_{1}^{r_n}{}_k\, a_{nk} \frac{e^{i\lambda_k t}}{i\lambda_k},$$

that is,

$$G_n'(t) = f_n(t).$$

Since

$$\lim_{n\to\infty} G_n(t) = G(t),$$

and

$$\lim_{n\to\infty} G_n'(t) = \lim_{n\to\infty} f_n(t) = f(t),$$

uniformly, relation (6.5) is then proved.

(c) *Let us prove theorem V*. Let $\varphi(\lambda) \in C^3(J)$, with

$$\varphi(\lambda) = \begin{cases} 1 & \text{when} \quad 0 \le |\lambda| < 1 \\ 0 & \text{when} \quad |\lambda| \ge 2. \end{cases}$$

Let

$$\psi(s) = \frac{1}{2\pi}\int_J \varphi(\lambda)e^{i\lambda s}d\lambda = \frac{1}{2\pi}\int_{-2}^{2} \varphi(\lambda)e^{i\lambda s}d\lambda \tag{6.6}$$

be the inverse Fourier transform of $\varphi(\lambda)$; $\psi(s)$ is obviously an entire function of s. Integrating twice by parts, we have, when $s\neq 0$,

$$|\psi(s)| = \left|\frac{1}{2\pi}\int_J \varphi''(\lambda)\frac{e^{i\lambda s}}{s^2}\,d\lambda\right| \le \frac{1}{2\pi s^2}\int_{-2}^{2} |\varphi''(\lambda)|d\lambda,$$

that is, $\psi(s) \in L^1(J)$.

Integrating by parts three times, we obtain, when $s\neq 0$,

$$|\psi'(s)| = \left|\frac{1}{2\pi}\int_J \lambda\varphi(\lambda)e^{i\lambda s}d\lambda\right| = \left|\frac{1}{2\pi}\int_J (\lambda(\varphi(\lambda))''' \frac{e^{i\lambda s}}{s^3}\,d\lambda\right|$$
$$\le \frac{1}{2\pi|s|^3}\int_{-2}^{2} |(\lambda\varphi(\lambda))'''|d\lambda.$$

Hence,

$$|s\psi'(s)| \le \frac{K}{1+s^2} \qquad (K = \text{const.}), \tag{6.7}$$

that is, $s\psi'(s) \in L^1(J)$.

Let us now set

(6.8)
$$\varphi_n(\lambda) = \varphi(n\lambda) \qquad (n = 1, 2, \cdots),$$
$$\psi_n(s) = \frac{1}{2\pi}\int_J \varphi_n(\lambda)e^{i\lambda s}d\lambda = \frac{1}{2\pi}\int_J \varphi(n\lambda)e^{i\lambda s}d\lambda,$$

from which follows

(6.9)
$$\psi_n(s) = \frac{1}{n}\psi\left(\frac{s}{n}\right), \qquad \psi_n'(s) = \frac{1}{n^2}\psi'\left(\frac{s}{n}\right).$$

Hence, by (6.6), (6.8), and (6.9) and by the inversion formula of the Fourier transform,

(6.10)
$$\varphi_n(\lambda) = \int_J \psi_n(s)e^{-i\lambda s}ds \Rightarrow \int_J \psi_n(s)ds = \varphi_n(0) = 1.$$

Set

$$f_n^{(1)}(t) = f * \psi_n(t),$$
$$f_n^{(2)}(t) = f(t) - f_n^{(1)}(t).$$

As $\psi_n(t) \in L^1(J)$, $f_n^{(1)}(t)$ is a.p. and, by the second of (6.9), admits the Fourier expansion

$$f_n^{(1)}(t) \sim \sum_{1}^{\infty}{}_k\, a_k\varphi_n(\lambda_k)e^{i\lambda_k t} \sim \sum_{1}^{\infty}{}_k\, a_k\varphi(n\lambda_k)e^{i\lambda_k t}.$$

Hence, $f_n^{(2)}(t)$ is also a.p. and has the Fourier expansion

$$f_n^{(2)}(t) \sim \sum_{1}^{\infty}{}_k\, a_k(1-\varphi(n\lambda_k))e^{i\lambda_k t},$$

so that the coefficients in the expansion of $f_n^{(2)}(t)$ vanish when $|\lambda_k| \leq 1/n$.

Setting

$$F(t) = F_n^{(1)}(t) + F_n^{(2)}(t) = \int_0^t f_n^{(1)}(\eta)d\eta + \int_0^t f_n^{(2)}(\eta)d\eta,$$

it follows by Favard's theorem that $F_n^{(2)}(t)$ is a.p., $\forall n$.

The theorem will therefore be proved provided we show that

$$\lim_{n\to\infty} F_n^{(1)}(t) = 0$$

uniformly.

We have, by (1.9) and the second of (6.10),

$$\begin{aligned} F_n^{(1)}(t) &= \int_0^t d\eta \int_J f(\eta-s)\psi_n(s)ds = \int_J \psi_n(s)ds \int_0^t f(\eta-s)d\eta \\ &= \int_J \psi_n(s)(F(t-s)-F(-s))ds \\ &= \int_J \psi_n(s)(F(t-s)-b)ds - \int_J \psi_n(s)(F(-s)-b)ds. \end{aligned}$$

Hence, it will be sufficient to prove that

$$\lim_{n\to\infty} \int_J \psi_n(s)(F(t-s)-b)ds = 0 \tag{6.11}$$

uniformly.

Integrating by parts, we obtain

$$\int_J (F(t-s)-b)\psi_n(s)ds = -\int_J\left\{\int_0^s (F(t-\sigma)-b)d\sigma\right\}\psi_n'(s)ds. \tag{6.12}$$

For every $\varepsilon>0$ there exists, by (1.9), $T_\varepsilon>0$ such that, when $|s|\geq T_\varepsilon$ and $\forall t\in J$,

$$\left\|\int_0^s (F(t-\sigma)-b)d\sigma\right\| \leq \varepsilon|s|. \tag{6.13}$$

Moreover, since $\|F(t)\|\leq M$ on J, then $\|b\|\leq M$ and, $\forall t\in J$,

$$\left\|\int_0^s (F(t-\sigma)-b)d\sigma\right\| \leq 2M|s|. \tag{6.14}$$

It follows, by (6.12), (6.13), and (6.14), that

$$\begin{aligned}\left\|\int_J (F(t-s)-b)\psi_n(s)ds\right\| &\leq 2M\int_{-T_\varepsilon}^{T_\epsilon} |s\psi_n'(s)|ds+\varepsilon\left\{\int_{T_\varepsilon}^{\infty} |s\psi_n'(s)|ds+\int_{-\infty}^{-T_\epsilon} |s\psi_n'(s)|ds\right\}\\ &\leq 2M\int_{-T_\varepsilon}^{T_\epsilon} |s\psi_n'(s)|ds+\varepsilon\int_J |s\psi_n'(s)|ds.\end{aligned} \tag{6.15}$$

Moreover, by the second of (6.8) and (6.9), we obtain

$$\int_{-T_\varepsilon}^{T_\epsilon} |s\psi_n'(s)|ds \leq \frac{1}{n^2}\int_{-T_\varepsilon}^{T_\epsilon} |s|ds\,\frac{1}{2\pi}\int_{-2}^{2} |\lambda\varphi(\lambda)|d\lambda \to 0, \tag{6.16}$$

and, by (6.9) and (6.7),

$$\int_J |s\psi_n'(s)|ds = \int_J \left|\frac{s}{n^2}\psi'\left(\frac{s}{n}\right)\right|ds = \int_J |s\psi'(s)|ds \leq K\int_J \frac{ds}{1+s^2} = K\pi. \tag{6.17}$$

From (6.15), (6.16), and (6.17) follows (6.11).

7. A.P. FUNCTIONS IN THE SENSE OF STEPANOV

(*a*) *Definition and some properties.* Let $x=f(t)\in L^p_{loc}(J;X)$, with $1\leq p<+\infty$; in other words, assume that, $\forall t\in J$ (with the exception at most of a set of measure zero), $f(t)$ takes its values in X, is integrable in the

Bochner-Lebesgue sense and the pth power of the norm $\|f(t)\|$ is Lebesgue-integrable on any bounded interval $a \vdash\!\dashv b$: $\int_a^b \|f(t)\|^p dt < +\infty$. Denoting by Δ the interval $0 \vdash\!\dashv 1$, this is obviously equivalent to assuming that

$$\left\{\int_\Delta \|f(t+\eta)\|^p d\eta\right\}^{1/p} < +\infty, \qquad \forall t \in J.$$

The function $f(t)$ is said to be almost-periodic in the sense of Stepanov (S^p a.p., or X-S^p a.p.) if to every $\varepsilon > 0$ there corresponds an r.d. set $\{\tau\}_\varepsilon$ such that, $\forall \tau \in \{\tau\}_\varepsilon$, we have

$$\text{(7.1)} \qquad \sup_J \left\{\int_\Delta \|f(t+\tau+\eta) - f(t+\eta)\|^p d\eta\right\}^{1/p} \leq \varepsilon.$$

It is clear that $f(t)$ *a.p.* $\Rightarrow f(t)$ S^p *a.p.*; moreover, *if* $p > m \geq 1$, then $f(t)$ S^p *a.p.* $\Rightarrow f(t)$ S^m *a.p.*

As has been observed by Bochner [1], *the almost-periodicity in the sense of Stepanov can be reduced to that in the sense of Bohr* (for *vector valued* functions).

Consider, in fact, the Banach space $L^p(\Delta; X)$ of functions $z(\eta)$, $\eta \in \Delta$, which take their values in X, almost everywhere (a.e.) on Δ, and such that $\int_\Delta \|z(\eta)\|^p d\eta < +\infty$. Since Δ is fixed, we shall denote $L^p(\Delta; X)$ by the more simple notation $L^p(X)$ and choose obviously the norm

$$\|z\|_{L^p(X)} = \left\{\int_\Delta \|z(\eta)\|_X^p d\eta\right\}^{1/p}.$$

This implies that two functions $z_1(\eta)$, $z_2(\eta)$ have to be considered identical (that is, $\|z_1 - z_2\|_{L^p(X)} = 0$) if they differ on a set of measure zero.

Suppose now that $g(t)$ is a function from J to $L^p(X)$; this means that $g(t) = \{g(t,\eta);\ \eta \in \Delta\}$ and

$$\|g(t)\| = \left\{\int_\Delta \|g(t,\eta)\|^p d\eta\right\}^{1/p}.$$

In particular, $g(t)$, continuous, is a.p. if, $\forall \varepsilon > 0$, $\exists \{\tau\}_\varepsilon$ r.d. such that, $\forall \tau \in \{\tau\}_\varepsilon$,

$$\text{(7.2)} \quad \sup_J \|g(t+\tau) - g(t)\| = \sup_J \left\{\int_\Delta \|g(t+\tau, \eta) - g(t,\eta)\|^p d\eta\right\}^{1/p} \leq \varepsilon.$$

Let $x = f(t) \in L^p_{loc}(J; X)$ and set, $\forall t \in J$,

$$\text{(7.3)} \qquad \tilde{f}(t) = \{f(t+\eta);\ \eta \in \Delta\}.$$

We obtain, in this way, a function from J to $L^p(X)$; moreover, since $f(t) \in L^p_{loc}(J; X)$, *the function $\tilde{f}(t)$*, with values in $L^p(X)$, *is continuous*, that is,

$$\text{(7.4)} \quad \lim_{\tau \to 0} \|\tilde{f}(t+\tau) - \tilde{f}(t)\| = \lim_{\tau \to 0} \left\{\int_\Delta \|f(t+\tau+\eta) - f(t+\eta)\|^p d\eta\right\}^{1/p} = 0.$$

The same $\tilde{f}(t)$ *is then a.p. if* (7.2) *holds*, that is, if, $\forall \tau \in \{\tau\}_\varepsilon$ r.d. ($\forall \varepsilon > 0$),

$$\sup_J \|\tilde{f}(t+\tau)-\tilde{f}(t)\| \leq \varepsilon. \tag{7.5}$$

By (7.1), (7.4), and (7.5), *the X-almost-periodicity of* $f(t)$ *in the sense of Stepanov is equivalent to the* $L^p(X)$*-almost-periodicity of* $\tilde{f}(t)$. This enables us to deduce, from the theory already developed, the theory of S^p a.p. functions.

We shall say, consequently, that $f(t)$ *is* S^p *weakly almost-periodic* (S^p *w.a.p.*, or X-S^p *w.a.p.*) *if the function* $\tilde{f}(t)$ *is w.a.p.*

For this, we recall that, if q is the conjugate exponent of p ($(1/p)+(1/q)=1$ when $p>1$, $q=+\infty$ when $p=1$), we obtain $(L^p(X))^*= L^q(X^*)$; hence (denoting, for clarity, by ${}_{Y^*}\langle\cdot,\cdot\rangle_Y$ the *duality* between a space Y and its dual space Y^*),

$${}_{L^q(X^*)}\langle z^*, \tilde{f}(t)\rangle_{L^p(X)} = \int_\Delta {}_{X^*}\langle z^*(\eta), f(t+\eta)\rangle_X d\eta,$$

where $z^*(\eta)$ is an arbitrary function with values in X^* (a.e. on Δ) and

$$\|z^*\|_{L^q(X^*)} = \left\{\int_\Delta \|z^*(\eta)\|^q_{X^*} d\eta\right\}^{1/q} \qquad \text{when} \quad p > 1$$

$$\|z^*\|_{L^q(X^*)} = \operatorname*{Ess\,sup}_\Delta \|z^*(\eta)\|_{X^*} \qquad \text{when} \quad p = 1.$$

Let us add now some properties of S^p a.p. functions.

VII (*Bochner*). $f(t)$ S^p*a.p. and* $f(t)$ *u.c.* $\Rightarrow f(t)$ *a.p.*

Let us prove, first, that, *if* $\varphi(t)$ *is a numerical continuous function with compact support, the convolution*

$$f * \varphi(t) = \int_J f(t-\eta)\varphi(\eta)d\eta = \int_J f(\eta)\varphi(t-\eta)d\eta \tag{7.6}$$

is X-a.p.

After having eventually performed a linear transformation on the independent variables ($t=as+b$, $a\neq 0$, $s\in J$), we may assume that $\operatorname{supp}\varphi(t)\subseteq\Delta$.

Setting $\rho=\max_\Delta |\varphi(t)|$, we have

$$\|f * \varphi(t+\tau)-f * \varphi(t)\| \leq \int_\Delta |\varphi(\eta)|\,\|f(t+\tau-\eta)-f(t-\eta)\|d\eta$$

$$\leq \rho \int_\Delta \|f(t+\tau-\eta)-f(t-\eta)\|d\eta.$$

Since $f(t)$ is S^1 a.p., the thesis is proved.

Let us now choose $\varphi(t)$ in such a way that

$$\varphi(t)\begin{cases} \geq 0 & \text{when} \quad t\in\Delta \\ = 0 & \text{when} \quad t\notin\Delta \end{cases}, \quad \int_\Delta \varphi(t)dt = 1.$$

We also set $\varphi_n(t) = n\varphi(nt)$. Hence,

$$\varphi_n(t) \geq 0, \qquad \text{supp}\, \varphi_n(t) \subseteq 0 \vdash\!\!\dashv \frac{1}{n}, \quad \int_0^{1/n} \varphi_n(t)dt = 1.$$

As $f(t)$ is u.c., there exists, $\forall \varepsilon > 0$, $\delta_\varepsilon > 0$ such that

$$\|f(t'') - f(t')\| \leq \varepsilon$$

when $|t' - t''| \leq \delta_\varepsilon$.

Taking n_ε so that $(1/2n_\varepsilon) \leq \delta_\varepsilon$, we have, when $n \geq n_\varepsilon$,

$$\|f * \varphi_n(t) - f(t)\| = \left\| n \int_J \varphi(n\eta)(f(t-\eta) - f(t))d\eta \right\|$$
$$\leq n \int_0^{1/n} \varphi(n\eta) \|f(t-\eta) - f(t)\| d\eta \leq \varepsilon.$$

Hence, $f(t)$, being the limit of a uniformly convergent sequence of a.p. functions, is a.p.

VIII. $f(t)$ *S^p a.p.*, $F(t) = \int_0^t f(\eta)d\eta \Rightarrow F(t)$ *u.c.*

Let us take $p = 1$ and assume that $F(t)$ is not u.c. There exist then $\rho > 0$ and two sequences $\{\sigma_n\}$ and $\{\delta_n\}$, with $\lim_{n\to\infty} \delta_n = 0$, such that

$$\|F(\sigma_n + \delta_n) - F(\sigma_n)\| = \left\| \int_0^{\delta_n} f(\sigma_n + \eta)d\eta \right\| \geq \rho. \tag{7.7}$$

As $f(t)$ is S^1 a.p., we may (by Bochner's criterion) assume that it is uniformly

$$\lim_{n\to\infty} \int_\Delta \|f(t + \sigma_n + \eta) - f_\sigma(t+\eta)\| d\eta = 0,$$

$f_\sigma(t)$ being, like $f(t)$, S^1 a.p. Moreover,

$$\left\| \int_0^{\delta_n} f(\sigma_n + \eta)d\eta \right\| \leq \left\| \int_0^{\delta_n} f_\sigma(\eta)d\eta \right\| + \left\| \int_0^{\delta_n} (f(\sigma_n + \eta) - f_\sigma(\eta))d\eta \right\| \to 0,$$

which contradicts (7.7). Hence $F(t)$ is u.c.

(*b*) *Harmonic analysis.*

IX. *Let $f(t)$ be S^p a.p. The Fourier expansion then holds*

$$f(t) \sim \sum_{1}^{\infty} {}_n\, a_n e^{i\lambda_n t}, \tag{7.8}$$

where

$$a_n = \lim_{T\to\infty} \frac{1}{T} \int_\sigma^{\sigma+T} f(t) e^{-i\lambda_n t} dt \tag{7.9}$$

uniformly with respect to σ.

Setting $\tilde{f}(t)=\{f(t+\eta);\ \eta \in \Delta\}$, $\tilde{f}(t)$ is, by assumption, $L^p(X)$-a.p. and the Fourier expansion

$$\tilde{f}(t) \sim \sum_{1}^{\infty}{}_n\, b_n e^{i\lambda_n t} \tag{7.10}$$

holds, where $b_n=\{b_n(\eta);\ \eta \in \Delta\} \in L^p(X)$. Moreover, we obtain, uniformly with respect to $\sigma \in J$,

$$b_n = \lim_{T\to\infty} \frac{1}{T}\int_{\sigma}^{\sigma+T} \tilde{f}(t)e^{-i\lambda_n t}dt. \tag{7.11}$$

Now, by (7.3), $\tilde{f}(t)e^{-i\lambda_n t}=\{f(t+\eta)e^{-i\lambda_n t};\ \eta \in \Delta\}$ and

$$\frac{1}{T}\int_{\sigma}^{\sigma+T} \tilde{f}(t)e^{-i\lambda_n t}dt = \left\{\frac{1}{T}\int_{\sigma}^{\sigma+T} f(t+\eta)e^{-i\lambda_n t}dt;\ \eta \in \Delta\right\}.$$

Relation (7.11) can then be written

$$\lim_{T\to\infty}\int_{\Delta}\left\| b_n(\eta)-\frac{1}{T}\int_{\sigma}^{\sigma+T} f(t+\eta)e^{-i\lambda_n t}dt\right\|^p d\eta = 0$$

uniformly with respect to σ. Hence,

$$\lim_{T\to\infty}\int_{\Delta}\left\| b_n(\eta)e^{-i\lambda_n\eta}-\frac{1}{T}\int_{\sigma+\eta}^{\sigma+T+\eta} f(t)e^{-i\lambda_n t}dt\right\|^p d\eta = 0 \tag{7.12}$$

uniformly.

As the function $f(t)e^{-i\lambda_n t}$ is S^p a.p., the integral

$$\int_0^t f(t)e^{-i\lambda_n t}dt$$

is u.c., $\forall n$. Hence,

$$\left\|\int_{\sigma}^{\sigma+\eta} f(t)e^{-i\lambda_n t}dt\right\| \leq m_n,$$

m_n being a constant which does not depend on $\sigma \in J$ and $\eta \in \Delta$.

By (7.12), we have then

$$\lim_{T\to\infty}\int_{\Delta}\left\| b_n(\eta)e^{-i\lambda_n\eta}-\frac{1}{T}\int_{\sigma}^{\sigma+T} f(t)e^{-i\lambda_n t}dt\right\|^p d\eta = 0 \tag{7.13}$$

uniformly with respect to σ. Setting $\sigma=0$, we obtain, by the theorem of Fischer-Riesz, a.e. on Δ,

$$\lim_{k\to\infty}\frac{1}{T_k}\int_0^{T_k} f(t)e^{-i\lambda_n t}dt = b_n(\eta)e^{-i\lambda_n\eta}, \tag{7.14}$$

where $\{T_k\}$ is a suitable sequence (depending on n), with $\lim_{k\to\infty} T_k=\infty$.

From (7.14), it follows that $b_n(\eta)e^{-i\lambda_n\eta}$ does not depend on η, that is,

$$b_n(\eta)e^{-i\lambda_n\eta} = a_n \in X. \tag{7.15}$$

Hence, by (7.13), we obtain, uniformly with respect to σ,

$$\lim_{T\to\infty} \frac{1}{T}\int_{\sigma}^{\sigma+T} f(t)e^{-i\lambda_n t}dt = a_n,$$

which proves (7.9). By (7.13) and (7.15), relation (7.10) can be written

$$\tilde{f}(t) \sim \left\{\sum_{1}^{\infty}{}_n\, b_n(\eta)e^{i\lambda_n t};\ \eta\in\Delta\right\} = \left\{\sum_{1}^{\infty}{}_n\, a_n e^{i\lambda_n(t+\eta)};\ \eta\in\Delta\right\} \tag{7.16}$$
$$\sim \{f(t+\eta);\ \eta\in\Delta\},$$

which justifies (7.8).

Regarding the summability of the series on the right-hand side of (7.8), observe that, applying Bochner's summation method to the expansion of $\tilde{f}(t)$, we have, by (7.16) and (3.17) of Chapter 2, uniformly

$$\lim_{m\to\infty}\left\|\tilde{f}(t)-\sum_{1}^{r_m}{}_k\,\mu_{mk}b_k e^{i\lambda_k t}\right\|_{L^p(X)} = 0. \tag{7.17}$$

In (7.17) the constants μ_{mk} are convergence factors, depending only on the sequence $\{\lambda_n\}$; moreover, (7.17) is equivalent to

$$\lim_{m\to\infty}\left\{\int_{\Delta}\left\|f(t+\eta)-\sum_{1}^{r_m}{}_k\,\mu_{mk}a_k e^{i\lambda_k(t+\eta)}\right\|^p d\eta\right\}^{1/p} = 0. \tag{7.18}$$

Hence *the approximation theorem for S^p a.p. functions is proved.*

By the convention of summing the series on the right-hand side in the way indicated by (7.18), (7.8) can be written

$$f(t) = \sum_{1}^{\infty}{}_n\, a_n e^{i\lambda_n t}. \tag{7.19}$$

(c) *Integration.* We shall now prove some results that, being valid also for S^p a.p. functions, enable us to reduce the hypothesis of compactness, or boundedness, of the integral $F(t)$ which was assumed in the preceding paragraphs.

Setting

$$\tilde{F}(t) = \{F(t+\eta);\ \eta\in\Delta\}, \tag{7.20}$$

we shall restrict ourselves to the generalization of theorems I, II, and V.

I′. *$f(t)$ S^p a.p., $\mathscr{R}_{\tilde{F}(t)}$ r.c. $\Rightarrow$ $F(t)$ a.p.*

We begin by observing that

$$\int_0^t \tilde{f}(s)ds = \left\{\int_0^t f(s+\eta)ds;\ \eta\in\Delta\right\} \tag{7.21}$$
$$= \left\{\int_0^{t+\eta} f(s)ds - \int_0^{\eta} f(s)ds;\ \eta\in\Delta\right\} = \tilde{F}(t)-\tilde{F}(0).$$

As, by assumption, $\tilde{f}(t)$ is a.p. and $\tilde{F}(t)$ has r.c. range, it follows that $\tilde{F}(t)$ is a.p. Hence $F(t)$ is S^p a.p. By theorem VIII, $F(t)$ is, on the other hand, u.c. and therefore, by theorem VII, $F(t)$ is a.p.

II′ (*Vasconi* [1], *Prouse* [2]). *X uniformly convex, $p>1$, $f(t)$ S^p a.p., $\tilde{F}(t)$ bounded $\Rightarrow$ $F(t)$ a.p.*

By assumption,

$$\sup_J \|\tilde{F}(t)\| = \sup_J \left\{\int_\Delta \|F(t+\eta)\|^p d\eta\right\}^{1/p} < +\infty. \tag{7.22}$$

Recall now that, if X is uniformly convex and $p>1$, the space $L^p(X)$ is uniformly convex. Hence, by theorem II and (7.21), $\tilde{F}(t)$ is a.p. Repeating the procedure given above, we conclude that also $F(t)$ is a.p.

V′ (*Levitan* [2]). *Let $f(t)$ be S^p a.p., $\tilde{F}(t)$ bounded and*

$$\lim_{T\to\infty} \frac{1}{T}\int_\sigma^{\sigma+T} F(t)dt = b \tag{7.23}$$

uniformly with respect to σ. Then $F(t)$ is a.p.

Setting $\tilde{b}=\{b;\ \eta\in\Delta\}$, we have, in fact,

$$\left\|\frac{1}{T}\int_\sigma^{\sigma+T} \tilde{F}(s)ds-\tilde{b}\right\| = \left\{\int_\Delta \left\|\frac{1}{T}\int_\sigma^{\sigma+T} F(s+\eta)ds-b\right\|^p d\eta\right\}^{1/p}$$
$$= \left\{\int_\Delta \left\|\frac{1}{T}\int_{\sigma+\eta}^{\sigma+T+\eta} F(s)ds-b\right\|^p d\eta\right\}^{1/p} \to 0$$

uniformly with respect to σ, because of (7.23).

It follows, by theorem V, that $\tilde{F}(t)$ is a.p. Hence $F(t)$ is S^p a.p. Since $F(t)$ is, by theorem VIII, u.c., the almost-periodicity of $F(t)$ follows then from theorem VII.

Observation IV. To simplify the notations in the following chapters, when $f(t)\in L^p_{loc}(J;X)$, $p\geq 1$, we shall write $f(t)$ instead of $\tilde{f}(t)$, adding the indication of the space in which the function has to be considered. We shall therefore write $f(t)=\{f(t+\eta);\ \eta\in\Delta\}$ and, by (7.4), $f(t)$ is $L^p(X)$-continuous. Therefore:

$$f(t)\in L^p_{loc}(J;X) \Rightarrow f(t)\ L^p(X)\text{-continuous;}$$

(7.24) $f(t)$ S^p *a.p.* (or S^p *w.a.p.*) *means* $f(t)$ $L^p(X)$-*a.p.* (or $L^p(X)$-*w.a.p.*);

$$\sup_J \left\{\int_\Delta \|f(t+\eta)\|^p d\eta\right\}^{1/p} < +\infty \text{ means } f(t)\ L^p(X)\text{-bounded.}$$

Authors' Remark. The open question mentioned at §1 has now been affirmatively solved by M. I. Kadets (On the integration of a.p. functions in a Banach space, *Funct. An. and Appl.* **3**, 1969).

PART II

APPLICATIONS TO ALMOST-PERIODIC FUNCTIONAL EQUATIONS

CHAPTER 5

THE WAVE EQUATION

1. THE INITIAL-BOUNDARY VALUE PROBLEM: EXISTENCE, UNIQUENESS, AND CONTINUOUS DEPENDENCE THEOREMS

(*a*) Let

$$a_0 x^{(n)}(t) + a_1 x^{(n-1)}(t) + \cdots + a_n x(t) = f(t) \tag{a$_1$}$$

be an ordinary linear differential equation with constant coefficients.

In the theory of numerical a.p. functions the following statement (theorem of Bohr-Neugebauer) holds: *if $f(t)$ is a.p. and $x(t)$ is a bounded solution of* (a$_1$), *then $x(t)$ is a.p.*

Obviously, this statement contains as a particular case the theorem of Bohl-Bohr.

We shall prove (in § 4) that an analogous result holds for a classical partial differential equation, with coefficients *independent* of t: the *wave equation*. The concepts of boundedness and almost-periodicity will naturally be intended in relation to the functional spaces (of fundamental physical interest) in which the solutions and the known term are defined.

(*b*) It is advisable, for the sake of clarity, to begin by studying the *wave equation* (or *equation of the vibrating membrane*):

$$\frac{\partial^2 x(t,\zeta)}{\partial t^2} = \sum_{j,k}^{1\cdots m} \frac{\partial}{\partial \zeta_j}\left(a_{jk}(\zeta)\,\frac{\partial x(t,\zeta)}{\partial \zeta_k}\right) - a_0(\zeta)x(t,\zeta) + f(t,\zeta). \tag{1.1}$$

More precisely, we shall consider the first initial-boundary value problem (or Cauchy-Dirichlet problem).

Let Ω be an *open*, *bounded*, and *connected* set of the Euclidean space R^m, $\partial\Omega$ the boundary of Ω, $\zeta = \{\zeta_1, \cdots, \zeta_m\}$ an arbitrary point of R^m.

The problem consists in finding a solution $x = x(t,\zeta)$ satisfying the *initial conditions*

$$x(0,\zeta) = x_0(\zeta), \qquad x_t(0,\zeta) = y_0(\zeta) \qquad (\zeta \in \Omega) \tag{1.2}$$

(with $x_0(\zeta)$, $y_0(\zeta)$ given functions) and the *boundary condition*

$$x(t,\zeta)\big|_{\zeta\in\partial\Omega} = 0. \tag{1.3}$$

This problem corresponds therefore to the study of the motion of a *vibrating membrane, with fixed edge.*

We shall assume that the functions considered in (1.1) are *real*.

It is common practice to consider, in the theory of partial differential equations, the so-called *weak* or *generalized* solutions. In the present case these are the solutions associated with the *variational* rather than the *differential* theory of the vibrating membrane; when some appropriate smoothness conditions are satisfied, the weak solutions verify, in the classical sense, (1.1), (1.2), and (1.3).

Let us first define the functional spaces in which the problem turns out to be *well posed*. As we shall see, these will be Hilbert spaces, as is often the case in the mathematical description of physical problems. The physical phenomena are, in fact, essentially bound to the concept of *energy* and therefore of *work*—hence, to that of *scalar product*. Now Hilbert spaces are just those Banach spaces in which a scalar product is defined, with the same *formal properties* that this functional has in Euclidean spaces.

We shall assume that the coefficients $a_{jk}(\zeta)$, $a_0(\zeta)$ are *measurable* and *bounded* functions on Ω, and that, $\forall \xi \in R^m$ and $\zeta \in \Omega$,

$$a_{jk}(\zeta) = a_{kj}(\zeta), \qquad \sum_{j,k}^{1\cdots m} a_{jk}(\zeta)\xi_j\xi_k \geq \nu \sum_{1}^{m}{}_j\, \xi_j^2 \quad (\nu > 0), \tag{1.4}$$
$$a_0(\zeta) \geq 0.$$

Hypotheses (1.4) are standard in the theory of elliptic equations (to which the hyperbolic equation (1.1) reduces when f and x do not depend on t): if no hypotheses of differentiability are made on the coefficients $a_{jk}(\zeta)$, then (1.1) must be interpreted in the sense of the theory of distributions.

Consider now the following Hilbert spaces.

1. The space $L^2(\Omega) = L^2$ of functions $x = \{x(\zeta);\ \zeta \in \Omega\}$ square summable on Ω, with the usual definition of scalar product (and hence of norm)

$$(x_1, x_2)_{L^2} = \int_\Omega x_1(\zeta)x_2(\zeta)d\zeta, \qquad \|x\|_{L^2} = (x,x)_{L^2}^{1/2}. \tag{1.5}$$

2. The space $H_0^1(\Omega) = H_0^1$ of functions that are square summable on Ω, together with their first derivatives, and vanish on $\partial\Omega$. The derivatives must be intended in the *generalized sense of Sobolev or of the theory of distributions* (that is, $g(\zeta) = \partial x/\partial \zeta_k$ means that $g \in L^2$ and that

$$\int_\Omega x(\zeta)\frac{\partial h(\zeta)}{\partial \zeta_k}\, d\zeta = -\int_\Omega g(\zeta)h(\zeta)d\zeta,$$

$\forall h \in \mathscr{D}(\Omega)$, set of all functions continuous on Ω, together with all their partial derivatives and with compact support on Ω). Moreover the vanishing of $x(\zeta)$ on $\partial\Omega$ must be intended (as will be hereafter explained) in the sense of the variational theory of elliptic equations.

The scalar product and the norm in H_0^1 can be defined in the following way:

$$(1.6)\quad (x_1,x_2)_{H_0^1} = \int_\Omega \left\{ \sum_{j,k}^{1\ldots m} a_{jk}(\zeta)\,\frac{\partial x_1(\zeta)}{\partial\zeta_j}\,\frac{\partial x_2(\zeta)}{\partial\zeta_k} + a_0(\zeta)x_1(\zeta)x_2(\zeta)\right\} d\zeta,$$

$$\|x\|_{H_0^1} = (x,x)_{H_0^1}^{1/2}.$$

We recall that the space H_0^1 can be obtained as the closure of $\mathscr{D}(\Omega)$ in the norm defined above. We have $H_0^1 \subset L^2$; the *embedding* of H_0^1 in L^2 is, moreover, not only *continuous* ($\|x\|_{L^2} \le \tilde{k}\|x\|_{H_0^1}$, $\tilde{k}>0$ embedding constant), but also *compact* (or *completely continuous*); this means that every sequence $\{x_n\}$, bounded in H_0^1, contains a subsequence that converges in L^2.

3. The space $E(\Omega)=E=H_0^1\times L^2$, topological product of H_0^1 by L^2. Each element $z\in E$ corresponds to a pair $\{x,y\}$, with $x\in H_0^1$, $y\in L^2$, and

$$(1.7)\qquad (z_1,z_2)_E = (x_1,x_2)_{H_0^1} + (y_1,y_2)_{L^2}, \qquad \|z\|_E = \{\|x\|_{H_0^1}^2 + \|y\|_{L^2}^2\}^{1/2}.$$

We shall call E the *energy space*, and the metric defined by the second of (1.7) the *energy metric*.

Assume now that the known term $f(t,\zeta)$ and the unknown function $x(t,\zeta)$ satisfy the following conditions.

(i_1) Setting $f(t)=\{f(t,\zeta);\ \zeta\in\Omega\}$, $f(t)$ takes its values in L^2 a.e. on J and its norm is locally summable, i.e., we obtain, $\forall$ bounded interval $a \vdash\!\dashv b$,

$$\int_a^b \|f(t)\|_{L^2}dt = \int_a^b \left\{\int_\Omega f^2(t,\zeta)d\zeta\right\}^{1/2} dt < +\infty.$$

In other words,

$$(1.8)\qquad f(t)\in L^1_{\text{loc}}(J;\ L^2).$$

(i_2) If we set $x(t)=\{x(t,\zeta);\ \zeta\in\Omega\}$, $x(t)$ takes its values in H_0^1, $\forall t\in J$, and is continuous, i.e.,

$$\lim_{\tau\to 0} \|x(t+\tau)-x(t)\|_{H_0^1}$$

$$= \lim_{\tau\to 0}\left\{\int_\Omega \left(\sum_{j,k}^{1\ldots m} a_{jk}(\zeta)\,\frac{\partial(x(t+\tau,\zeta)-x(t,\zeta))}{\partial\zeta_j}\right.\right.$$

$$\left.\left. \times\,\frac{\partial(x(t+\tau,\zeta)-x(t,\zeta))}{\partial\zeta_k} + a_0(\zeta)(x(t+\tau,\zeta)-x(t,\zeta))^2\right)d\zeta\right\}^{1/2} = 0.$$

We shall moreover assume that $x(t)$, from J to L^2, is *differentiable* and that its derivative $x'(t)=\{\partial x(t,\zeta)/\partial t;\ \zeta \in \Omega\}$ (in the *strong* sense, i.e.,

$$\lim_{\tau\to 0}\left\|\frac{x(t+\tau)-x(t)}{\tau}-x'(t)\right\|_{L^2}=0)$$

is *continuous* on J. Hence, if we set $z(t)=\{x(t),x'(t)\}$, the function $z(t)$ is *continuous* from J to E:

$$(1.9)\qquad x(t)\in C^0(J;\,H_0^1),\qquad x'(t)\in C^0(J;\,L^2),\qquad z(t)\in C^0(J;\,E).$$

It is important to observe that the quantities

$$(1.10)\qquad \begin{gathered}\tfrac{1}{2}\|x(t)\|^2_{H_0^1},\qquad \tfrac{1}{2}\|x'(t)\|^2_{L^2},\\ \tfrac{1}{2}\|z(t)\|^2_E=\tfrac{1}{2}\|x(t)\|^2_{H_0^1}+\tfrac{1}{2}\|x'(t)\|^2_{L^2}\end{gathered}$$

measure, respectively, the *potential energy*, the *kinetic energy*, and the *total energy*, at the time t, of the vibrating membrane. The denomination of *energy space* given to E is therefore justified.

In what follows, in order not to complicate the notation, we shall again denote by $x(t)$ the pair $\{x(t),x'(t)\}$, writing $x(t)=\{x(t),x'(t)\}$, or also, more precisely, $x(t)\underset{E}{=}\{x(t),x'(t)\}$; hence the last of (1.10) is equivalent to setting

$$(1.11)\qquad \|x(t)\|_E=\{\|x(t)\|^2_{H_0^1}+\|x'(t)\|^2_{L^2}\}^{1/2}.$$

We recall that, *by Hamilton's principle, the functions* $x(t)$, $t\in J$, *which describe the possible motions of the membrane, are those for which the integral*

$$\mathscr{H}(x)=\int_a^b\{\tfrac{1}{2}\|x'(t)\|^2_{L^2}-\tfrac{1}{2}\|x(t)\|^2_{H_0^1}+(f(t),x(t))_{L^2}\}dt$$

(*Hamiltonian action*) *is stationary*, $\forall$ *interval* $a\vdash\!\dashv b$, *with respect to all the variations* $h(t)$ *of the same functional class as* $x(t)$ *and with support* $\subset a\vdash\!\dashv b$.

By imposing that the first variation $\delta\mathscr{H}(x)$ of the functional $\mathscr{H}(x)$ vanish, we obtain the *variational wave equation*

$$(1.12)\qquad \int_J\{(x'(t),h'(t))_{L^2}-(x(t),h(t))_{H_0^1}+(f(t),h(t))_{L^2}\}dt=0,$$

which must be satisfied $\forall h(t)\in C^0(J;H_0^1)$, with compact support, and $h'(t)\in C^0(J;\,L^2)$.

In the present chapter we shall make frequent reference to equation (1.12).

It is well known that, by means of Green's formula, (1.12) can be obtained from (1.1) and (1.3).

If, moreover, $x(t)$ satisfies (1.12), condition (1.3) is (by the definition itself of H_0^1) verified, whereas, if $x(t,\zeta)$, $\partial\Omega$ and the coefficients $a_{jk}(\zeta)$ satisfy obvious smoothness conditions, then, being $h(t,\zeta)$ arbitrary, it is possible to see that $x(t,\zeta)$ is a solution of (1.1).

For this reason, *the solutions $x(t)$ of the variational wave equation are called weak, or generalized, solutions of the differential wave equation.*

(c) Let us consider, for (1.12), the *initial value problem*, or *Cauchy problem* (to which problem (1.1), (1.2), (1.3) reduces itself, having assumed that $x(t)$ takes its values in H_0^1). We wish to *find a solution $x(t)$, $t \in J$, satisfying the initial conditions*:

$$x(0) = x_0, \qquad x'(0) = y_0 \qquad (\Leftrightarrow x(0) \underset{E}{=} \{x_0, y_0\}), \tag{1.13}$$

x_0 and y_0 being arbitrary functions of H_0^1 and L^2, respectively.

Let us prove the following theorem (see Nagy [1], Ladyženskaja [1], Amerio [2]).

I. *If $f(t) \in L^1_{\text{loc}}(J; L^2)$, the initial value problem admits one and only one solution, for any choice of $x_0 \in H_0^1$ and $y_0 \in L^2$. Moreover, the following evaluation holds*

$$\|x(t)\|_E \leq \|x(0)\|_E + \left| \int_0^t \|f(\eta)\|_{L^2} d\eta \right|, \tag{1.14}$$

i.e., the solution $x(t)$ depends continuously on the initial value $x(0)$ and on the known term $f(t)$.

To prove the thesis, we shall use the method of elementary solutions (or Fourier's method).

Consider the sequence $\{w_n\}$ of the *eigensolutions* of the equation

$$(w,l)_{H_0^1} = \lambda (w,l)_{L^2}, \tag{1.15}$$

which must be satisfied $\forall l \in H_0^1$. As is well known from the variational theory of elliptic equations, (1.15) admits a sequence $\{\lambda_n\}$ of eigenvalues, such that

$$0 < \lambda_1 \leq \lambda_2 \leq \cdots \leq \lambda_n \leq \cdots, \qquad \lim_{n \to \infty} \lambda_n = +\infty.$$

In what follows we shall set

$$\gamma_n = \sqrt{\lambda_n}.$$

Correspondingly, the sequence $\{w_n\}$ of *eigensolutions* satisfies the orthonormality conditions:

$$\left(\frac{w_n}{\gamma_n}, \frac{w_m}{\gamma_m} \right)_{H_0^1} = (w_n, w_m)_{L^2} = \delta_{nm}. \tag{1.16}$$

The sequence $\{w_n\}$ is moreover complete, both in L^2 and in H_0^1.

Let us assume that problem (1.12), (1.13) has a solution and calculate it. Expanding in Fourier series, we have

$$x(t) \underset{H_0^1}{=} \sum_1^\infty {}_n\, \alpha_n(t) \frac{w_n}{\gamma_n} \underset{L^2}{=} \sum_1^\infty {}_n\, \frac{\alpha_n(t)}{\gamma_n} w_n \tag{1.17}$$

where, by (1.15) and (1.16),

$$\alpha_n(t) = \left(x(t), \frac{w_n}{\gamma_n}\right)_{H_0^1} = \gamma_n(x(t),w_n)_{L^2}. \tag{1.18}$$

Hence,

$$\alpha_n'(t) = \gamma_n(x'(t),w_n)_{L^2} \tag{1.19}$$

and, consequently,

$$x'(t) \underset{L^2}{=} \sum_1^\infty{}_n \frac{\alpha_n'(t)}{\gamma_n} w_n. \tag{1.20}$$

On the other hand,

$$f(t) \underset{L^2}{=} \sum_1^\infty{}_n \varphi_n(t)w_n \qquad (\varphi_n(t) = (f(t),w_n)_{L^2}). \tag{1.21}$$

Let us now set, in (1.12),

$$h(t) = \theta(t)\frac{w_j}{\gamma_j} \qquad (j = 1, 2, \cdots),$$

$\theta(t)$ being a continuous function, together with its derivative $\theta'(t)$, and with compact support. From (1.17), (1.20), and (1.21), it follows, by (1.16), that

$$(x(t),h(t))_{H_0^1} = \theta(t)\alpha_j(t), \qquad (x'(t),h'(t))_{L^2} = \theta'(t)\frac{\alpha_j'(t)}{\gamma_j^2},$$

$$(f(t),h(t))_{L^2} = \theta(t)\frac{\varphi_j(t)}{\gamma_j},$$

so that (1.12) becomes

$$\int_J \theta'(t)\frac{\alpha_j'(t)}{\gamma_j^2}\,dt = \int_J \theta(t)\left(\alpha_j(t) - \frac{\varphi_j(t)}{\gamma_j}\right)dt.$$

Since $\theta(t)$ is arbitrary, there exists then the derivative $\alpha_j''(t)$ and we have

$$\frac{\alpha_j''(t)}{\gamma_j^2} = -\alpha_j(t) + \frac{\varphi_j(t)}{\gamma_j}. \tag{1.22}$$

It follows that

$$\alpha_j(t) = C_j \cos\gamma_j t + D_j \sin\gamma_j t + \int_0^t \varphi_j(\eta)\sin\gamma_j(t-\eta)\,d\eta, \tag{1.23}$$

where the constants C_j, D_j are determined by means of the initial conditions (1.13); precisely, by (1.18), (1.19),

$$C_j = \left(x_0, \frac{w_j}{\gamma_j}\right)_{H_0^1}, \qquad D_j = (y_0,w_j)_{L^2}. \tag{1.24}$$

If the solution $x(t)$ exists, it is therefore unique.

Let us now prove that the series (1.17) and (1.20) converge uniformly on every bounded interval (so that $x(t)$ and $x'(t)$ will be, respectively, H_0^1 and L^2-continuous; in other words, $x(t)$ will be E-continuous).

Consider, for $1 \le p \le q$, the function, from J to E,

$$x_{pq}(t) = \left\{ \sum_p^q {}_n\, \alpha_n(t) \frac{w_n}{\gamma_n}, \sum_p^q {}_n\, \frac{\alpha_n'(t)}{\gamma_n} w_n \right\} \tag{1.25}$$

and observe that, by (1.16) and (1.23), we obtain

$$\begin{aligned} \|x_{pq}(t)\|_E &= \left\{ \sum_p^q {}_n \left[\left(C_n \cos \gamma_n t + D_n \sin \gamma_n t + \int_0^t \varphi_n(\eta) \sin \gamma_n (t-\eta) d\eta \right)^2 \right.\right. \\ &\qquad + \left(-C_n \sin \gamma_n t + D_n \cos \gamma_n t \right. \\ &\qquad\qquad \left.\left.\left. + \int_0^t \varphi_n(\eta) \cos \gamma_n (t-\eta) d\eta \right)^2 \right] \right\}^{1/2} \\ &= \left\{ \sum_p^q {}_n \left[\left(C_n - \int_0^t \varphi_n(\eta) \sin \gamma_n \eta d\eta \right)^2 \right.\right. \\ &\qquad\qquad \left.\left. + \left(D_n + \int_0^t \varphi_n(\eta) \cos \gamma_n \eta d\eta \right)^2 \right] \right\}^{1/2} \\ &= \left\{ \sum_p^q {}_n\, |(D_n + iC_n) + \int_0^t \varphi_n(\eta) e^{-i\gamma_n \eta} d\eta|^2 \right\}^{1/2} \\ &\le \left\{ \sum_p^q {}_n\, |D_n + iC_n|^2 \right\}^{1/2} + \left\{ \sum_p^q {}_n \left| \int_0^t \varphi_n(\eta) e^{-i\gamma_n \eta} d\eta \right|^2 \right\}^{1/2}. \end{aligned} \tag{1.26}$$

Since

$$\|x(0)\|_E^2 = \|x(0)\|_{H_0^1}^2 + \|x'(0)\|_{L^2}^2 = \sum_1^\infty {}_n\, (C_n^2 + D_n^2) < +\infty, \tag{1.27}$$

we have

$$\lim_{(p,q)\to\infty} \sum_p^q {}_n\, |D_n + iC_n|^2 = 0. \tag{1.28}$$

On the other hand,

$$\begin{aligned} \sum_p^q {}_n \left| \int_0^t \varphi_n(\eta) e^{-i\gamma_n \eta} d\eta \right|^2 &\le \sum_p^q {}_n \left(\int_0^t |\varphi_n(\eta)| d\eta \right)^2 \\ &= \left\| \int_0^t g_{pq}(\eta) d\eta \right\|_{L^2}^2 \le \left| \int_0^t \|g_{pq}(\eta)\|_{L^2} d\eta \right|^2, \end{aligned} \tag{1.29}$$

where

$$g_{pq}(t) \underset{L^2}{=} \sum_p^q {}_n\, |\varphi_n(t)| w_n.$$

Setting

$$(1.30)\qquad f_{pq}(t) = \sum_{p}^{q}{}_{n}\, \varphi_n(t)w_n, \qquad f_p(t) = \sum_{p}^{\infty}{}_{n}\, \varphi_n(t)w_n,$$

we have $\|g_{pq}(t)\|_{L^2} = \|f_{pq}(t)\|_{L^2}$. Hence, by (1.29) and (1.30),

$$(1.31)\quad \sum_{p}^{q}{}_{n} \left|\int_0^t \varphi_n(\eta)e^{-i\gamma_n\eta}d\eta\right|^2 \leq \left|\int_0^t \|g_{pq}(\eta)\|_{L^2}d\eta\right|^2 \leq \left|\int_0^t \|f_p(\eta)\|_{L^2}d\eta\right|^2.$$

Since, by the second of (1.30),

$$\lim_{p\to\infty} \|f_p(t)\|_{L^2} = 0$$

a.e. on J and, moreover, $\|f_p(t)\|_{L^2} \leq \|f(t)\|_{L^2}$, we obtain (by a known theorem of Lebesgue on integration by series)

$$(1.32)\qquad \lim_{p\to\infty} \int_0^t \|f_p(\eta)\|_{L^2}d\eta = 0$$

uniformly on every bounded interval.

From (1.26), (1.28), and (1.32) it follows that the series (1.17) and (1.20) converge uniformly on every bounded interval; $x(t)$ is therefore E-continuous and the initial conditions (1.13) are satisfied.

Finally, it is easy to see that (1.12) is satisfied ∀ variation $h(t)$ of the kind previously defined.

(d) The *evaluation* (1.14) follows directly from the inequality (deducted from (1.25), (1.26), and (1.31)):

$$\|x_{pq}(t)\|_E \leq \|x_{pq}(0)\|_E + \left|\int_0^t \|f_p(\eta)\|_{L^2}d\eta\right|.$$

It is, in fact, sufficient to set $p=1$ and let q diverge.

2. ENERGY EQUATION. HOMOGENEOUS EQUATION AND ENERGY CONSERVATION PRINCIPLE

(a) Let $x_1(t)$ and $x_2(t)$ be two solutions of the wave equation, corresponding to the known terms $f_1(t)$ and $f_2(t)$, respectively. The fundamental *energy equation* then holds

$$(2.1)\quad (x_1(t),x_2(t))_E = (x_1(0),x_2(0))_E + \int_0^t \{(x_1'(\eta),f_2(\eta))_{L^2} + (x_2'(\eta),f_1(\eta))_{L^2}\}d\eta.$$

Denoting, in fact, by $\alpha_{nj}(t)$, $\varphi_{nj}(t)$ the Fourier coefficients corresponding to $x_j(t)$, $f_j(t)$ we have, by (1.22),

$$\begin{aligned}\frac{\alpha'_{n1}(t)}{\gamma_n}\,\varphi_{n2}(t)+\frac{\alpha'_{n2}(t)}{\gamma_n}\,\varphi_{n1}(t) &= \frac{\alpha'_{n1}(t)\alpha''_{n2}(t)}{\gamma_n^2}+\frac{\alpha'_{n2}(t)\alpha''_{n1}(t)}{\gamma_n^2}+\alpha'_{n1}(t)\alpha_{n2}(t)+\alpha'_{n2}(t)\alpha_{n1}(t)\\ &= \frac{d}{dt}\left(\alpha_{n1}(t)\alpha_{n2}(t)+\frac{\alpha'_{n1}(t)}{\gamma_n}\,\frac{\alpha'_{n2}(t)}{\gamma_n}\right),\end{aligned}$$

that is,

$$\begin{aligned}(2.2)\quad \alpha_{n1}(t)\alpha_{n2}(t)+\frac{\alpha'_{n1}(t)}{\gamma_n}\,\frac{\alpha'_{n2}(t)}{\gamma_n} &= \alpha_{n1}(0)\alpha_{n2}(0)+\frac{\alpha'_{n1}(0)}{\gamma_n}\,\frac{\alpha'_{n2}(0)}{\gamma_n}+\int_0^t\left\{\frac{\alpha'_{n1}(\eta)}{\gamma_n}\,\varphi_{n2}(\eta)+\frac{\alpha'_{n2}(\eta)}{\gamma_n}\,\varphi_{n1}(\eta)\right\}d\eta.\end{aligned}$$

It follows then, by (2.2), that

$$\begin{aligned}(2.3)\qquad (x_1(t),x_2(t))_E &= (x_1(0),x_2(0))_E\\ &\quad+\sum_1^\infty{}_n\int_0^t\left\{\frac{\alpha'_{n1}(\eta)}{\gamma_n}\,\varphi_{n2}(\eta)+\frac{\alpha'_{n2}(\eta)}{\gamma_n}\,\varphi_{n1}(\eta)\right\}d\eta.\end{aligned}$$

On the other hand,

$$(x'_1(t),f_2(t))_{L^2}+(x'_2(t),f_1(t))_{L^2} = \sum_1^\infty{}_n\frac{\alpha'_{n1}(t)}{\gamma_n}\,\varphi_{n2}(t)+\frac{\alpha'_{n2}(t)}{\gamma_n}\,\varphi_{n1}(t)$$

and, for $r=1, 2,\cdots$,

$$\begin{aligned}(2.4)\quad &\left|\sum_1^r{}_n\frac{\alpha'_{n1}(t)}{\gamma_n}\,\varphi_{n2}(t)+\frac{\alpha'_{n2}(t)}{\gamma_n}\,\varphi_{n1}(t)\right|\\ &\le\left\{\sum_1^r{}_n\frac{\alpha'^2_{n1}(t)}{\gamma_n^2}\right\}^{1/2}\left\{\sum_1^r{}_n\varphi_{n2}^2(t)\right\}^{1/2}+\left\{\sum_1^r{}_n\frac{\alpha'^2_{n2}(t)}{\gamma_n^2}\right\}^{1/2}\left\{\sum_1^r{}_n\varphi_{n1}^2(t)\right\}^{1/2}\\ &\le \|x'_1(t)\|_{L^2}\|f_2(t)\|_{L^2}+\|x'_2(t)\|_{L^2}\|f_1(t)\|_{L^2},\end{aligned}$$

which, by (1.8) and the second of (1.9), is a locally summable function. We can then apply to the right-hand side of (2.3) Lebesgue's theorem on integration by series, thus proving (2.1).

(*b*) We now consider the *homogeneous wave equation*

$$(2.5)\qquad \int_J\{(u'(t),h'(t))_{L^2}-(u(t),h(t))_{H_0^1}\}dt = 0,$$

where, for the sake of clarity, we have denoted by $u(t)$ the unknown function.

From (2.1) we obtain, setting $f(t)\equiv 0$,

$$(2.6)\qquad (u_1(t),u_2(t))_E = (u_1(0),u_2(0))_E,$$

i.e., the scalar product of two solutions of the homogeneous equation is constant.

In particular, setting $u_1(t)=u_2(t)=u(t)$ we obtain, by (2.6),

$$\|u(t)\|_E = \|u(0)\|_E, \tag{2.7}$$

i.e., the solutions of the homogeneous equation have constant norm in the energy space; the ranges are, in other words, *spherical lines, with centers at the origin.*

Equation (2.7) expresses the *energy conservation principle.*

Observe, lastly, that if $u(t)$ is a solution of (2.5), $u(t+\tau)-u(t)$ also is, $\forall\tau\in J$, a solution. Hence, by (2.7),

$$\|u(t+\tau)-u(t)\|_E = \|u(\tau)-u(0)\|_E, \tag{2.8}$$

i.e., the principle of conservation of distances holds (see Chapter 1, § 2).

3. A MINIMAX THEOREM

Assume that (1.12) admits a solution, $x_1(t)$, with *bounded energy,* i.e., such that

$$\sup_J \|x_1(t)\|_E < +\infty.$$

As every other solution, $x(t)$, is obtained by adding to $\bar{x}_1(t)$ a solution, $u(t)$, of the homogeneous equation, it follows from (2.7) that $x(t)$ also is E-bounded.

Let us now set

$$\mu(x) = \sup_J \|x(t)\|_E \tag{3.1}$$

and prove the following statement, which we shall call the *minimax theorem* (Amerio [2]).

II. *Assume that there exists an E-bounded solution, $x_1(t)$, and let Γ be the set of all the solutions. Setting*

$$\tilde{\mu} = \inf_\Gamma \mu(x), \tag{3.2}$$

there exists one and only one solution, $\tilde{x}(t)$, such that

$$\mu(\tilde{x}) = \tilde{\mu}. \tag{3.3}$$

We shall call $\tilde{x}(t)$ the *minimal* solution of (1.12).

Hence, *among all the solutions of the variational wave equation there is one, and only one, for which the supremum, on J, of the energy takes on the smallest possible value.*

We shall deduce the theorem from the following *lemma.*

Let Z be a uniformly convex space and G a bounded set $\subset Z$. Setting

$$\rho(z) = \sup_{g\in G} \|z-g\| \qquad (\forall z \in Z), \tag{3.4}$$

$$\tilde{\rho} = \inf_{z\in Z} \rho(z), \tag{3.5}$$

there exists one and only one point, $\tilde{z}$, such that

$$\rho(\tilde{z}) = \tilde{\rho}. \tag{3.6}$$

Observe, first of all, that $\rho(z)$ is the radius of the smallest closed circular neighborhood of the point z, containing G; $\tilde{z}$ and $\tilde{\rho}$ define therefore the smallest closed circle that contains G. For this reason, $\tilde{z}$ and $\tilde{\rho}$ will be called, respectively, the *center* and the *radius* of the set G.

Let $\{z_n\}$ be a minimizing sequence for the functional $\rho(z)$, i.e., such that

$$\tilde{\rho} \leq \rho(z_n) = \tilde{\rho}+\varepsilon_n, \qquad 0 \leq \varepsilon_n \leq 1, \qquad \lim_{n\to\infty} \varepsilon_n = 0. \tag{3.7}$$

We shall prove that $\{z_n\}$ is a Cauchy sequence.

If not, there would exist a constant $\sigma>0$ and two subsequences $\{z_{n1}\}\subset\{z_n\}$, $\{z_{n2}\}\subset\{z_n\}$ such that

$$\|z_{n1}-z_{n2}\| \geq \sigma.$$

Hence, $\forall g \in G$ (being $\|z_n-g\| \leq \rho(z_n) \leq \tilde{\rho}+1$),

$$\|(z_{n1}-g)-(z_{n2}-g)\| \geq \sigma \geq \frac{\sigma}{\tilde{\rho}+1} \max\{\|z_{n1}-g\|, \|z_{n2}-g\|\}.$$

Setting $\sigma/(\tilde{\rho}+1)=\theta>0$, we have therefore (by (4.28) and (4.29), Chapter 3)

$$\begin{aligned}\left\|\frac{z_{n1}+z_{n2}}{2}-g\right\| &\leq (1-\omega(\theta))\max\{\|z_{n1}-g\|, \|z_{n2}-g\|\}\\ &\leq (1-\omega(\theta))\,(\tilde{\rho}+\varepsilon_{n1}+\varepsilon_{n2}).\end{aligned}$$

It follows, by (3.4), taking $n \geq \bar{n}$ sufficiently large, that

$$\rho\left(\frac{z_{n1}+z_{n2}}{2}\right) \leq (1-\omega(\theta))\,(\tilde{\rho}+\varepsilon_{n1}+\varepsilon_{n2}) < \tilde{\rho}.$$

Since this is absurd, the limit

$$\lim_{n\to\infty} z_n = \tilde{z}$$

exists and we have, $\forall g \in G$,

$$\|\tilde{z}-g\| = \lim_{n\to\infty} \|z_n-g\| \leq \lim_{n\to\infty} \rho(z_n) = \tilde{\rho} \Rightarrow \rho(\tilde{z}) = \tilde{\rho}.$$

Assume now that it is also $g(w)=g(\tilde{z})$. We have then, necessarily, $w=\tilde{z}$, since $\tilde{z}, w, \tilde{z}, w, \cdots$ is a minimizing sequence for the functional $g(z)$.

Let us now prove the minimax theorem.

We begin by observing that, by (1.26),

$$\|x(t)\|_E^2 = \sum_1^\infty {}_n \left(C_n - \int_0^t \varphi_n(\eta) \sin \gamma_n \eta d\eta\right)^2 + \left(D_n + \int_0^t \varphi_n(\eta) \cos \gamma_n \eta d\eta\right)^2 \tag{3.8}$$

which gives

$$\sum_1^\infty {}_n (C_n^2 + D_n^2) = \|x(0)\|_E^2 < +\infty. \tag{3.9}$$

Consider the Hilbert space $Z=l^2$ of all real sequences

$$z = \{C_1, D_1, C_2, D_2, \cdots, C_n, D_n, \cdots\}$$

such that

$$\|z\|_{l^2} = \left\{\sum_1^\infty {}_n (C_n^2 + D_n^2)\right\}^{1/2} < +\infty.$$

Let G be, in l^2, the line of parametric equation

$$g = g(t) = \left\{\int_0^t \varphi_1(\eta) \sin \gamma_1 \eta d\eta, \; -\int_0^t \varphi_1(\eta) \cos \gamma_1 \eta d\eta, \cdots, \right.$$
$$\left. \int_0^t \varphi_n(\eta) \sin \gamma_n \eta d\eta, \; -\int_0^t \varphi_n(\eta) \cos \gamma_n \eta d\eta, \cdots \right\}.$$

By assumption, the set G is bounded and (3.8) can be written

$$\|x(t)\|_E^2 = \|z - g(t)\|_{l^2}^2.$$

Hence, by (3.1) and (3.4),

$$\mu(x) = \sup_J \|x(t)\|_E = \sup_J \|z - g(t)\|_{l^2} = \rho(z)$$

and, by the lemma we have just proved, there exists one and only one sequence

$$\tilde{z} = \{\tilde{C}_n, \tilde{D}_n\}$$

such that

$$\rho(\tilde{z}) = \inf_{z \in l^2} \rho(z) = \tilde{\rho}.$$

Relation (3.3) is therefore proved: precisely,

$$\tilde{\mu} = \tilde{\rho} \tag{3.10}$$

and

$$\tilde{x}(t) \underset{H_0^1}{=} \sum_1^\infty {}_n \left\{\tilde{C}_n \cos \gamma_n t + \tilde{D}_n \sin \gamma_n t + \int_0^t \varphi_n(\eta) \sin \gamma_n (t-\eta) d\eta\right\} \frac{w_n}{\gamma_n}. \tag{3.11}$$

Corollary. Assume that $f(t)$ is *periodic, with period* Θ; assume moreover that *there exists an E-bounded solution*, $x_1(t)$. Under these assumptions, *the minimal solution* $\tilde{x}(t)$ *is periodic with period* Θ.

The proof follows immediately, from the uniqueness of the minimal solution. In fact, if $f(t)$ is periodic with period Θ, then $\tilde{x}(t)$ and $\tilde{x}(t+\Theta)$ are minimal solutions; hence $\tilde{x}(t)=\tilde{x}(t+\Theta)$.

In this way we have proved, for (1.12), an *existence theorem of periodic solutions*. The periodic solution $\tilde{x}(t)$ is obtained by the minimax theorem and not (as often occurs in dissipative systems) as asymptotic value, when $t \to +\infty$, of an arbitrary solution.

Finally, it is hardly necessary to recall that, if $f(t)$ is periodic, there may exist no bounded solutions; the *resonance* phenomenon may in fact occur. If, for example, $f(t)=(\cos \gamma_k t)w_k$, we have the unbounded solution $x(t)=(t \sin \gamma_k t/2\gamma_k)w_k$.

4. A.P. SOLUTIONS OF THE WAVE EQUATION WITH A.P. KNOWN TERM

(*a*) The E-almost-periodicity of the solutions $u(t)$ of the *homogeneous* equation (2.5) has been proved by many authors under always wider assumptions: Muckenhoupt [1], Bochner [2], Bochner and Von Neumann [2], Sobolev [2], Ladyženskaja [1].

Very significant is the *deduction* (due to Bochner) *of the almost-periodicity of* $u(t)$ *from the energy conservation principle under the assumption that the range* $\mathscr{R}_{u(t)}$ *be E-r.c.*

Equation (2.8), in fact, holds and, assuming that $\mathscr{R}_{u(t)}$ is E-r.c., the thesis follows by theorem X, Chapter 1.

Subsequently, Sobolev eliminated the compactness hypothesis, assuming that the boundary $\partial\Omega$ has continuous curvatures; finally Ladyženskaja eliminated also this hypothesis, thus obtaining the following statement.

III. *The solutions* $u(t)$ *of the homogeneous equation are E-a.p.*

Observe, in fact, that, by (1.17) and (1.23), we obtain

$$(4.1)\quad u(t) \underset{E}{=} \left\{\sum_1^\infty{}_n (C_n \cos \gamma_n t + D_n \sin \gamma_n t)\frac{w_n}{\gamma_n}, \right.$$

$$\left. \sum_1^\infty{}_n (-C_n \sin \gamma_n t + D_n \cos \gamma_n t) w_n \right\}$$

and the series of a.p. functions on the right-hand side of (4.1) E-converges uniformly. In fact, when $1 \le p \le q$,

$$\left\| \sum_{p}^{q} {}_n (C_n \cos \gamma_n t + D_n \sin \gamma_n t) \frac{w_n}{\gamma_n} \right\|_{H_0^1}^2 + \left\| \sum_{p}^{q} {}_n (-C_n \sin \gamma_n t + D_n \cos \gamma_n t) w_n \right\|_{L^2}^2$$

$$= \sum_{p}^{q} {}_n (C_n^2 + D_n^2) \to 0,$$

since, by (1.27),

$$\sum_{1}^{\infty} {}_n (C_n^2 + D_n^2) = \|u(0)\|_E^2 < +\infty.$$

(*b*) *Let us consider the nonhomogeneous wave equation* (1.12), *assuming that the known term is a.p. from J to L^2.*

In this case it is possible that no bounded solution exists; however, as we have already observed, if a bounded solution exists, all solutions are bounded.

Regarding almost-periodicity, it has been proved by Zaidman [1] that *if the range $\mathscr{R}_{x(t)}$ is E-r.c., then the solution $x(t)$ is E-a.p.*

Subsequently, Amerio [3], [6] eliminated the compactness hypothesis, substituting it by one of *boundedness*, which is strictly necessary and has an evident physical interpretation.

The following theorem therefore holds.

IV. *$f(t)$ L^2-a.p., $x(t)$ E-bounded $\Rightarrow$ $x(t)$ E-a.p.*

The proof can be given by the following methods, analogous to those of theorems III and II of Chapter 4.

(b_1) Observe that if $f(t)$ is L^2-a.p. and

$$\sup_J \|x(t)\|_E = M < +\infty, \tag{4.2}$$

the functions $\alpha_n(t)$ defined by (1.23) are a.p. together with their derivatives $\alpha_n'(t)$. The function $\varphi_n(t) = (f(t), w_n)_{L^2}$ is, in fact, a.p.; moreover, by (1.23) and (4.2),

$$\alpha_n^2(t) + \frac{\alpha_n'^2(t)}{\gamma_n^2} = \left(C_n - \int_0^t \varphi_n(\eta) \sin \gamma_n \eta d\eta \right)^2 + \left(D_n + \int_0^t \varphi_n(\eta) \cos \gamma_n \eta d\eta \right)^2 \le M^2, \tag{4.3}$$

so that the integrals

$$\int_0^t \varphi_n(\eta) \sin \gamma_n \eta d\eta, \qquad \int_0^t \varphi_n(\eta) \cos \gamma_n \eta d\eta$$

are bounded and therefore a.p.

Hence, if we show that the series of a.p. functions defining the solution $x(t)$:

$$x(t) \underset{E}{=} \left\{ \sum_1^\infty {}_n\, \alpha_n(t) \frac{w_n}{\gamma_n}, \sum_1^\infty {}_n \frac{\alpha_n'(t)}{\gamma_n} w_n \right\}, \tag{4.4}$$

converges uniformly, we will have proved the almost-periodicity of $x(t)$. This is equivalent to proving that the series of nonnegative a.p. functions

$$\sum_1^\infty {}_n\, \alpha_n^2(t) + \frac{\alpha_n'^2(t)}{\gamma_n^2} = \|x(t)\|_E^2 \tag{4.5}$$

converges uniformly.

We shall utilize the extension of Dini's theorem to a.p. functions (Chapter 4, theorem VI), as has already been done for the integration of a.p. functions.

Let us prove first that *the norm* $\|x(t)\|_E$ *is a.p.* In fact, by (2.1) (if we set $x_1(t) = x_2(t), f_1(t) = f_2(t)$),

$$\frac{d}{dt} \|x(t)\|_E^2 = 2(x'(t), f(t))_{L^2} = 2 \sum_1^\infty {}_n \frac{\alpha_n'(t)}{\gamma_n} \varphi_n(t). \tag{4.6}$$

Observe now that the series of a.p. functions

$$\sum_1^\infty {}_n \frac{\alpha_n'(t)}{\gamma_n} \varphi_n(t)$$

is uniformly convergent.

We have, in fact, for $1 \leq p \leq q$,

$$\begin{aligned} \left| \sum_p^q {}_n \frac{\alpha_n'(t)}{\gamma_n} \varphi_n(t) \right| &\leq \left\{ \sum_p^q {}_n \frac{\alpha_n'^2(t)}{\gamma_n^2} \right\}^{1/2} \left\{ \sum_p^q {}_n\, \varphi_n^2(t) \right\}^{1/2} \\ &\leq M \left\{ \sum_p^q {}_n\, \varphi_n^2(t) \right\}^{1/2} \end{aligned}$$

and, being $f(t)$ a.p., the series

$$\sum_1^\infty {}_n\, \varphi_n^2(t) = \|f(t)\|_{L^2}^2$$

converges uniformly (Chapter 3, theorem XIV).

By (4.6), the function $\|x(t)\|_E^2$ has, therefore, an a.p. derivative; moreover, $\|x(t)\|_E^2$ is bounded by (4.2). Hence $\|x(t)\|_E^2$ is a.p.

Now let $s=\{s_k\}$ be a regular sequence for the sequence $\{\alpha_n^2(t)+\alpha_n'^2(t)/\gamma_n^2\}$.

By the almost-periodicity, we can assume forthwith that, for any $n \geq 1$, we have

$$\begin{aligned} \lim_{k\to\infty} f(t+s_k) &= g(t), \\ \lim_{k\to\infty} \alpha_n(t+s_k) &= \delta_n(t), \\ \lim_{k\to\infty} \alpha_n'(t+s_k) &= \delta_n'(t), \end{aligned} \tag{4.7}$$

uniformly on J, the functions $g(t)$, $\delta_n(t)$, $\delta_n'(t)$ being (like $f(t)$, $\alpha_n(t)$, $\alpha_n'(t)$) a.p.

From (3.8) and (4.2), it follows that

$$\begin{aligned} \sum_1^\infty {}_n \left\{\alpha_n^2(t+s_k)+\frac{\alpha_n'^2(t+s_k)}{\gamma_n^2}\right\} &= \sum_1^\infty {}_n \left(\alpha_n(s_k)-\int_0^t \varphi_n(\eta+s_k)\sin\gamma_n\eta\,d\eta\right)^2 \\ &\quad + \left(\frac{\alpha_n'(s_k)}{\gamma_n}+\int_0^t \varphi_n(\eta+s_k)\cos\gamma_n\eta\,d\eta\right)^2 \\ &\leq M^2. \end{aligned} \tag{4.8}$$

Moreover, by (4.7), we have, $\forall n \geq 1$, uniformly on J,

$$\lim_{k\to\infty}\left(f(t+s_k), \frac{w_n}{\gamma_n}\right)_{H_0^1} = \left(g(t), \frac{w_n}{\gamma_n}\right)_{H_0^1} = \psi_n(t) \tag{4.9}$$

and ($\forall r \geq 1$; $t \in J$)

$$\begin{aligned} \sum_1^r {}_n \left\{\delta_n^2(t)+\frac{\delta_n'^2(t)}{\gamma_n^2}\right\} &= \sum_1^r {}_n \left(\delta_n(0)-\int_0^t \psi_n(\eta)\sin\gamma_n\eta\,d\eta\right)^2 \\ &\quad + \left(\frac{\delta_n'(0)}{\gamma_n}+\int_0^t \psi_n(\eta)\cos\gamma_n\eta\,d\eta\right)^2 \\ &\leq M^2. \end{aligned} \tag{4.10}$$

Therefore,

$$\sum_1^\infty {}_n\, \delta_n^2(t)+\frac{\delta_n'^2(t)}{\gamma_n^2} \leq M^2 \tag{4.11}$$

and, in particular, when $t=0$,

$$\sum_1^\infty {}_n\, \delta_n^2(0)+\frac{\delta_n'^2(0)}{\gamma_n^2} \leq M^2.$$

We conclude that the series

$$x_s(t) \underset{H_0^1}{=} \sum_1^\infty {}_n\, \delta_n(t)\, \frac{w_n}{\gamma_n} \tag{4.12}$$

defines the solution of the wave equation corresponding to the known term $g(t)$ and to the initial conditions

$$\left(x_s(0), \frac{w_n}{\gamma_n}\right)_{H_0^1} = \delta_n(0), \qquad (x_s'(0), w_n)_{L^2} = \delta_n'(0).$$

Therefore,

$$\|x_s(t)\|_E^2 = \sum_1^\infty {}_n\, \delta_n^2(t) + \frac{\delta_n'^2(t)}{\gamma_n^2}$$

and, by what has been proved above, $\|x_s(t)\|_E$ is an a.p. function, such being $g(t)$.

As this property holds $\forall$ regular sequence s, the uniform convergence of series (4.5) is then proved.

(b_2) The proof given in (b_1) associates the almost-periodicity of the solution $x(t)$ to the almost-periodicity of the Fourier coefficients $\alpha_n(t)$ and of the norms $\|x_s(t)\|_E$, $\forall s$; this last property follows from the energy relation.

The second proof of the almost-periodicity of $x(t)$ (which we shall now give) will be obtained by showing, first of all, that $x(t)$ is w.a.p. and subsequently that it has an r.c. range.

Let us prove that $x(t)$ is E-w.a.p.

Let

$$g = \left\{\sum_1^\infty {}_n\, \rho_n \frac{w_n}{\gamma_n}, \sum_1^\infty {}_n\, \sigma_n w_n\right\}$$

be an arbitrary element of E; therefore,

$$\|g\|_E = \left\{\sum_1^\infty {}_n\, \rho_n^2 + \sigma_n^2\right\}^{1/2} < +\infty \tag{4.13}$$

and, moreover,

$$(x(t), g)_E = \sum_1^\infty {}_n\, \alpha_n(t)\rho_n + \frac{\alpha_n'(t)}{\gamma_n}\,\sigma_n, \tag{4.14}$$

where $\alpha_n(t)$, $\alpha_n'(t)$ are a.p. In order to prove the thesis, it will be sufficient to show that the series on the right-hand side of (4.14) converges uniformly.

For $1 \leq p \leq q$, we obtain

$$\left|\sum_p^q {}_n\, \alpha_n(t)\rho_n + \frac{\alpha_n'(t)}{\gamma_n}\,\sigma_n\right| \leq \left\{\sum_p^q {}_n\, \alpha_n^2(t) + \frac{\alpha_n'^2(t)}{\gamma_n^2}\right\}^{1/2} \left\{\sum_p^q {}_n\, \rho_n^2 + \sigma_n^2\right\}^{1/2}$$

$$\leq M\left\{\sum_p^q {}_n\, \rho_n^2 + \sigma_n^2\right\}^{1/2}.$$

Hence, by (4.13), the uniform convergence is proved and $x(t)$ is *E-w.a.p.*

For proving the *E-almost periodicity* of $x(t)$ we may now utilize theorem XII of Chapter 3 (by which one proves that $x(t)$ *has an E-r.c. range*).

Assume that there exist a sequence $s=\{s_k\}$ and a number $\rho>0$ such that

$$\|x(s_j)-x(s_k)\|_E \geq \rho \qquad (j \neq k). \tag{4.15}$$

We may suppose ($f(t)$ being a.p.) that, uniformly,

$$\lim_{(j,k)\to\infty} (f(t+s_j)-f(t+s_k)) = 0. \tag{4.16}$$

To prove the thesis it will be sufficient to show that, $\forall t \in J$,

$$\max \lim_{(j,k)\to\infty} \|x(t+s_j)-x(t+s_k)\|_E \geq \rho. \tag{4.17}$$

Let us denote by $u_{jk}(t)$ the solution of the homogeneous equation corresponding to the initial conditions

$$u_{jk}(0) \underset{H_0^1}{=} x(s_j)-x(s_k), \qquad u'_{jk}(0) \underset{L^2}{=} x'(s_j)-x'(s_k) \tag{4.18}$$

and by $z_{jk}(t)$ the solution corresponding to the known term

$$f(t+s_j)-f(t+s_k)$$

and to the initial conditions

$$z_{jk}(0) \underset{H_0^1}{=} 0, \qquad z'_{jk}(0) \underset{L^2}{=} 0.$$

We have then

$$x(t+s_j)-x(t+s_k) = u_{jk}(t)+z_{jk}(t), \tag{4.19}$$

which implies that

$$\|x(t+s_j)-x(t+s_k)\|_E \geq \|u_{jk}(t)\|_E - \|z_{jk}(t)\|_E. \tag{4.20}$$

Moreover, by the energy conservation principle, the continuous dependence theorem, and by (4.15) and (4.16), we obtain

$$\|u_{jk}(t)\|_E = \|u_{jk}(0)\|_E \geq \rho,$$
$$\|z_{jk}(t)\|_E \leq \left|\int_0^t \|f(\eta+s_j)-f(\eta+s_k)\| d\eta\right| \to 0 \qquad (\forall t).$$

Hence (4.17) holds and the E-almost-periodicity of $x(t)$ is proved.

Observation. Generalizations and other results on the problem considered above have been given by Bochner [3], Zaidman [3], and Prouse [1]. For the C^0-almost-periodicity of the solutions, see Vaghi [1]. Vaghi [2] considers the generalization of theorem IV to the *weak solutions* (corresponding to problem (1.3)) *of the equation*

$$\frac{\partial^2 x}{\partial t^2}+\alpha(t)\frac{\partial x}{\partial t}+\beta(t)x = \sum_{j,k}^{1\ldots m}\frac{\partial}{\partial \zeta_j}\left(a_{jk}(\zeta)\frac{\partial x}{\partial \zeta_k}\right)+f(t,\zeta) \tag{4.21}$$

where $\alpha(t)$, $\beta(t)$ *are real periodic functions* $\in C^1(J)$ *with period* Θ, *and* $f(t)=\{f(t,\zeta);\ \zeta\in\Omega\}$ *is* L^2*-a.p.*

In particular, if $f(t)=0$, *the E-bounded eigensolutions of* (4.21) *are periodic or a.p.*

It may be noted that, although (4.21) is an equation with coefficients depending on t in a very special way, it represents the *variation equation* of an interesting equation.

Let

$$z''+g(z,z') = 0 \tag{4.22}$$

be, in fact, a nonlinear second-order equation, with $g(z,z')$ continuous together with its derivatives g_z, $g_{z'}$, for $-\infty<z,\ z'<+\infty$.

Assume that (4.22) admits a periodic solution $z=z_0(t)$, with period Θ. This function is also a solution, independent of ζ, of the nonlinear partial differential equation

$$\frac{\partial^2 z}{\partial t^2}+g\left(z,\frac{\partial z}{\partial t}\right) = \sum_{j,k}^{1\ldots m}\frac{\partial}{\partial \zeta_j}\left(a_{jk}(\zeta)\frac{\partial z}{\partial \zeta_k}\right). \tag{4.23}$$

Denoting by $x(t,\zeta)$ the variation given to $z_0(t)$, the variation equation corresponding to (4.23) is then

$$\frac{\partial^2 x}{\partial t^2}+g_z(z_0(t),z_0'(t))x+g_{z'}(z_0(t),z_0'(t))\frac{\partial x}{\partial t} = \sum_{j,k}^{1\ldots m}\frac{\partial}{\partial \zeta_j}\left(a_{jk}(\zeta)\frac{\partial x}{\partial \zeta_k}\right).$$

We obtain in this way a particular case of (4.21); by (1.3) we shall consider variations which vanish, $\forall t$, on the boundary $\partial\Omega$.

For equation (1.12), with $f(t)$ L^2-S^2 a.p., see Vasconi [2].

CHAPTER 6

THE SCHRÖDINGER TYPE EQUATION

1. INTRODUCTION AND STATEMENTS

(*a*) The essential aim of the present chapter is to extend to the equation of the Schrödinger type the fundamental results obtained by Favard [1] for the systems of linear ordinary differential equation, with a.p. coefficients and known term.

We shall, moreover, give an almost-periodicity theorem of the eigen-solutions of the same equation in the case when the operator $A(t)$, which appears in it, is *periodic*; we shall thus obtain a statement analogous to that which, for the homogeneous ordinary systems, with periodic coefficients, follows from a classical theorem of Liapunov.

Consider a linear ordinary system, which we shall write in the following vector form

$$(\mathrm{a}_1) \qquad ix'(t) = A(t)x(t)+f(t).$$

In equation (a_1) $x(t)$ and $f(t)$ are column vectors of the complex Euclidean space $\mathscr{C}^n$, $A(t)$ is a square complex matrix of order n, and $A(t)$ and $f(t)$ are a.p. functions.

Let $\mathscr{S}$ be the family of all the sequences $s=\{s_k\}$ *regular* with respect to $A(t)$, such that

$$\lim_{k\to\infty} A(t+s_k) = A_s(t)$$

uniformly. We obtain, in this way, a family of a.p. matrices $A_s(t)$; by Bochner's criterion, every real sequence $c=\{c_k\}$ contains a subsequence $s \in \mathscr{S}$.

In his theory, Favard considers, together with (a_1), the family of homogeneous equations:

$$(\mathrm{a}_2) \qquad iu'(t) = A_s(t)u(t) \qquad (A_0(t) = A(t))$$

and proves the following theorems.

(1) *Assume that*, $\forall s$, *equation* (a_2) *does not have bounded eigensolutions*.

Then, if there exists a (necessarily unique) *bounded solution*, $z(t)$, *of* (a_1), $z(t)$ *is a.p.*

(2) *Assume that,* $\forall s$, *any bounded eigensolution,* $u(t)$, *of* (a_2) *satisfies the condition:*

$$\inf_J \|u(t)\| > 0.$$

Then, if there exists a bounded solution, $z(t)$, *of* (a_1), *there exists also an a.p. solution,* $\tilde{x}(t)$.

Precisely, $\tilde{x}(t)$ *is the solution* (the existence and uniqueness of which are proved) *for which the functional*

$$\mu(x) = \sup_J \|x(t)\|$$

takes on its smallest value, when $x(t)$ *varies in the set of the bounded solutions of* (a_1).

Observe that, *in* (1), *the almost-periodicity theorem follows from a uniqueness theorem of the bounded solutions.*

It is clear that, in theorem (2), *Favard's condition*:

$$(a_3) \qquad u(t) \text{ bounded eigensolution of } (a_2) \Rightarrow \inf_J \|u(t)\| > 0,$$

imposes a strict limitation on the matrix $A(t)$. However, of interest is the fact that *Favard's condition is satisfied if* (a_2) *is self-adjoint,* i.e., if $A^*(t) = A(t)$ ($\Rightarrow A_s^*(t) = A_s(t)$). In fact, in this case, by (a_2),

$$(iu'(t),u(t)) = (A_s(t)u(t),u(t)),$$
$$(u(t),iu'(t)) = (u(t),A_s(t)u(t)) = (A_s(t)u(t),u(t)).$$

Hence,

$$(u'(t),u(t)) + (u(t),u'(t)) = 0,$$

that is, the known result:

$$\|u(t)\| = \text{const.}$$

Favard's theory can therefore be applied to self-adjoint systems: if (a_1) *(with* $A(t)$ *self-adjoint and a.p. matrix,* $f(t)$ *a.p.) admits a bounded solution,* $z(t)$, *it admits also one which is a.p.: the minimal solution* $\tilde{x}(t)$.

Regarding the nature of the eigensolutions $u(t)$ of (a_2) we note that, in general, these are *not* a.p., even if $A(t)$ is self-adjoint and a.p.

Consider, in fact, for $n = 1$, the equation

$$iu'(t) = \varphi(t)u(t),$$

with $\varphi(t)$ real a.p. function. We have

$$u(t) = u(0) \exp\left[i \int_0^t \varphi(\eta) d\eta\right]$$

and, by a theorem of Favard [1], the right-hand side is a.p. if and only if it is

$$(a_4) \qquad \int_0^t \varphi(\eta)d\eta = \alpha t + \psi(t),$$

with α constant, $\psi(t)$ a.p. function. If we then choose $\varphi(t)$, a.p., in such a way that (a_4) does not hold, $u(t)$ cannot be a.p.

We can, however, prove that *if* $A(t)$, *self-adjoint, is periodic, then all the eigensolutions* $u(t)$ *of* (a_2) *are a.p.*

We recall, in fact, that, by the theorem of Liapunov mentioned above, every periodic linear system can be *reduced*, by means of a linear periodic nonsingular transformation, to a system with constant coefficients. Under the assumptions made the system thus obtained has only bounded, and therefore a.p., solutions.

(*b*) Let X and Y be two complex Hilbert spaces. Assume $X \subseteq Y$, separable, dense in Y, with continuous embedding ($\|x\|_Y \leq \tilde{k}\|x\|$, $\tilde{k} > 0$, where $\|\cdot\|_Y$, $(\cdot,\cdot)_Y$ and $\|\cdot\|$, $(\cdot,\cdot)$ denote norm and scalar product on Y and X, respectively). Let $\mathscr{A} = \mathscr{L}(X,X)$ be the Banach space of the linear and bounded operators A, from X to X, with the norm corresponding to the *uniform topology*: $\|A\| = \sup_{\|x\|=1} \|Ax\|$.

We shall call "*Schrödinger type equation*" the equation:

$$(1.1) \qquad \int_J \{i(x(t),h'(t))_Y + (A(t)x(t) + f(t),h(t))\}dt = 0,$$

where the *unknown function* $x(t)$, the *operator* $A(t)$, the *known term* $f(t)$, and the *test function* $h(t)$ satisfy the following conditions:

(i_1) $x(t) \in C^0(J; X)$;

(i_2) $A(t) \in C^1(J; \mathscr{A})$ *is self-adjoint and satisfies the ellipticity condition*

$$(1.2) \qquad (A(t)x,x) \geq \nu\|x\|^2 \qquad (\nu > 0);$$

(i_3) $f(t) \in C^1(J; X)$;

(i_4) $h(t) \in C^0(J; X)$, $h'(t) \in C^0(J; Y)$. Moreover, $h(t)$ has *compact support* and (1.1) must be satisfied for *all* test functions $h(t)$.

We shall denote by $u(t)$ the solution of the *homogeneous equation*

$$(1.3) \qquad \int_J \{i(u(t),h'(t))_Y + (A(t)u(t),h(t))\}dt = 0.$$

Equation (1.1) is the weak form corresponding, for instance, to the following problem.

Let Ω be an open, connected, *bounded* or *unbounded* set of the Euclidean space R^m ($\zeta = \{\zeta_1, \cdots, \zeta_m\}$) and consider the equation

$$(1.4) \quad i\frac{\partial x(t,\zeta)}{\partial t} + \sum_{j,k}^{1\ldots m} \frac{\partial}{\partial \zeta_j}\left(a_{jk}(t,\zeta)\frac{\partial x(t,\zeta)}{\partial \zeta_k}\right) - a_0(t,\zeta)x(t,\zeta)$$
$$= \int_\Omega \Phi(t,\zeta,\xi)x(t,\xi)d\xi + f(t,\zeta) \qquad (t \in J,\ \zeta \in \Omega).$$

The derivatives will always be intended in the sense of the theory of distributions.

Denoting by $\bar{\alpha}$ the conjugate of the complex number α, we shall assume that the coefficients $a_{jk}(t,\zeta)$, $a_0(t,\zeta)$ are measurable and bounded functions on $J \times \Omega$ and that

$$(1.5) \qquad a_{jk}(t,\zeta) = \overline{a_{kj}(t,\zeta)},$$

$$\sum_{j,k}^{1\ldots m} a_{jk}(t,\zeta)\lambda_j\bar{\lambda}_k \geq \rho \sum_{1}^{m}{}_j |\lambda_j|^2 \qquad (\rho > 0), \quad a_0(t,\zeta) \geq \rho_0 \geq 0,$$

where $\rho_0 > 0$ if Ω is unbounded, $\rho_0 \geq 0$ if Ω is bounded. The second equation of (1.5) must hold for all complex values $\lambda_1, \cdots, \lambda_m$.

We shall moreover assume that the functions $a_{jk}(t,\zeta)$, $a_0(t,\zeta)$ and their first derivatives with respect to t are bounded on Ω, $\forall t \in J$, and that they are continuous functions of t, uniformly with respect to $\zeta \in \Omega$.

Suppose also that the *kernel* $\Phi(t,\zeta,\xi)$ is, $\forall t \in J$, positive semidefinite, self-adjoint, and $L^2(\Omega \times \Omega)$-continuous, together with its derivative $\Phi_t(t,\zeta,\xi)$.

Finally, we shall assume that the known term $f(t,\zeta)$ is $L^2(\Omega)$-continuous together with its derivative $f_t(t,\zeta)$.

The problem we shall consider consists in determining, on $J \times \Omega$, a solution $x(t,\zeta)$ satisfying the *initial condition*

$$(1.6) \qquad x(0,\zeta) = x_0(\zeta)$$

and the *boundary condition*

$$(1.7) \qquad x(t,\zeta)|_{\zeta \in \partial\Omega} = 0.$$

Let us now set

$$Y = L^2(\Omega), \qquad X = H_0^1(\Omega),$$

with the norms

$$\|x\| = \left\{\int_\Omega \left(\sum_{1}^{m}{}_j \left|\frac{\partial x(\zeta)}{\partial \zeta_j}\right|^2 + |x(\zeta)|^2\right) d\zeta\right\}^{1/2} = \left\{\sum_{1}^{m}{}_j \left\|\frac{\partial x}{\partial \zeta_j}\right\|_Y^2 + \|x\|_Y^2\right\}^{1/2}.$$

Hence the embedding of X in Y is continuous.

Moreover ($\forall x, z \in X$),

$$(A(t)x,z) = \int_\Omega \left\{\sum_{j,k}^{1\ldots m} a_{jk}(t,\zeta) \frac{\partial x(\zeta)}{\partial \zeta_k} \frac{\partial \bar{z}(\zeta)}{\partial \zeta_j} + a_0(t,\zeta)x(\zeta)\bar{z}(\zeta)\right\} d\zeta$$
$$+ \int_\Omega\int_\Omega \Phi(t,\zeta,\xi)x(\xi)\bar{z}(\zeta)\,d\zeta\,d\xi,$$

$$(A'(t)x,z) = \int_\Omega \left\{\sum_{j,k}^{1\ldots m} \frac{\partial a_{jk}(t,\zeta)}{\partial t} \frac{\partial x(\zeta)}{\partial \zeta_k} \frac{\partial \bar{z}(\zeta)}{\partial \zeta_j} + \frac{\partial a_0(t,\zeta)}{\partial t} x(\zeta)\bar{z}(\zeta)\right\} d\zeta$$
$$+ \int_\Omega\int_\Omega \Phi_t(t,\zeta,\xi)x(\xi)\bar{z}(\zeta)\,d\zeta\,d\xi$$

and (i_2) is therefore satisfied.

Setting $x(t)=\{x(t,\zeta);\ \zeta \in \Omega\}$, $f(t)=\{f(t,\zeta);\ \zeta \in \Omega\}$, we obtain, in the usual way, the weak form (1.1) of the problem; the solution $x(t)$ must satisfy the initial condition $x(0)=x_0=\{x_0(\zeta);\ \zeta \in \Omega\}$.

We now give the *fundamental equations* (which will be proved at § 2, *theorems I and III*) that hold for the solutions of (1.1) and (1.3):

$$\text{(a)}\quad \frac{d}{dt}\|x(t)\|_Y^2 = 2\mathscr{T}(f(t),x(t));$$

$$\text{(b)}\quad \frac{d}{dt}\{(A(t)x(t),x(t))+2\mathscr{R}(f(t),x(t))\} = (A'(t)x(t),x(t))+2\mathscr{R}(f'(t),x(t));$$

$$\text{(a}')\quad \|u(t)\|_Y = \|u(0)\|_Y;$$

$$\text{(b}')\quad (\textit{if } A(t) = I)\ \|u(t)\| = \|u(0)\|.$$

Equalities (a′) and (b′) show that *two principles of conservation of the norm* hold: *for the Y-norm the principle is always valid; for the X-norm it holds only when* $A=I$ (to this case we can, by (1.5), always reduce our problem if $A(t)=$const, i.e., for a_{jk}, a_0, Φ independent of t: we set, in fact,

$$(1.8)\qquad (x,z) = \int_\Omega \left(\sum_{j,k}^{1\ldots m} a_{jk}(\zeta)\frac{\partial x(\zeta)}{\partial \zeta_j}\frac{\partial \bar{z}(\zeta)}{\partial \zeta_k} + a_0(\zeta)x(\zeta)\bar{z}(\zeta)\right)d\zeta + \int_\Omega\int_\Omega \Phi(\zeta,\xi)x(\xi)\bar{z}(\zeta)d\zeta d\xi.$$

The initial value problem for (1.1), $x(0)=x_0$, *has*, $\forall x_0 \in X$, *one and only one solution* $x(t)$; *moreover*, $x(t)$ *depends continuously on* x_0 *and* $f(t)$; we have, in fact, $\forall$ interval $-T\le t\le T$,

$$(1.9)\qquad \|x(t)\| \le M_T\left\{\|x_0\| + \|f(0)\| + \int_{-T}^{T}\|f'(\eta)\|d\eta\right\},$$

where M_T does not depend on x_0 and $f(t)$ (see § 2, *theorem II*).

In the proofs we shall use the Faedo-Galerkin method and a "regularization" procedure (Lions and Prodi [1], Torelli [1], Strauss [1]. For the general theory of equation (1.1), see also Lions [2], and Lions and Magenes [1]).

As in the preceding chapters, when we shall say that a function $z(t)$ is bounded, or uniformly continuous (u.c.), or uniformly weakly continuous (u.w.c.), we shall always intend this for the whole interval J. Moreover, we shall add the notation of the space in which $z(t)$ takes its values with the only exception of the space X; hence $z(t)$ bounded, or u.c., or u.w.c., or a.p., or w.a.p. means that $z(t)$ is X-bounded, or X-u.c., or X-u.w.c., or X-a.p., or X-w.a.p.

We shall study, in §§ 3, 4, 5, equations (1.1) and (1.3) with the aim of giving conditions for the existence of *one* a.p. solution of (1.1), if $A(t)$ and $f(t)$ are a.p., and for the existence of a.p. eigensolutions of (1.3), if $A(t)$ is a.p. (Amerio [17]). In other words, we shall treat the extension to equations (1.1) and (1.3) of the results of Favard and Liapunov, recalled at (*a*).

First, we give the following *definitions*.

Let $z(t)$ be a bounded function and set

$$\mu(z) = \sup_J \|z(t)\|, \tag{1.10}$$

$$\varphi(z;\, v,\, \tau) = \sup_J |(z(t+\tau)-z(t),v)| \qquad (\forall v \in X,\ \tau \in J). \tag{1.11}$$

Let Γ_z be the set (obviously *convex*) of all solutions $x(t)$ of (1.1), bounded and such that

$$\varphi(x;\, v,\tau) \leq \varphi(z;\, v,\tau) \qquad (\forall v \in X,\ \tau \in J). \tag{1.12}$$

Obviously, the set Γ_z is not empty if equation (1.1) admits a bounded solution $z(t)$.

Let us now state the theorems IV, V, VI which correspond to the extension mentioned above.

IV (*Minimax theorem*). *Assume that there exists a bounded function, $z(t)$, such that the set Γ_z is not empty. Then, if*

$$\tilde{\mu} = \inf_{\Gamma_z} \mu(x), \tag{1.13}$$

there exists, in Γ_z, one and only one solution, $\tilde{x}(t)$, such that

$$\mu(\tilde{x}) = \tilde{\mu}. \tag{1.14}$$

$\tilde{x}(t)$ is therefore the *minimal* solution, in Γ_z.

It may be observed, comparing this theorem with the minimax theorem regarding the wave equation (Chapter 1, § 3) and with the theorem of Favard for ordinary systems, that the set Γ_z of the solutions is now more restricted, owing to condition (1.12). The reason for this is related to the hypothesis of uniform continuity for the bounded solution $z(t)$ (in theorems V and VI), a property which the minimal solution $\tilde{x}(t)$ must also have.

Corollary. *$A(t)$ and $f(t)$ periodic, with period Θ, $\Rightarrow$ $\tilde{x}(t)$ periodic with period Θ.*

V (*Almost-periodicity theorem*). *Assume that:*

(1) *The operators $A(t)$, $A'(t)$ are $\mathscr{A}$-a.p.;*

(2) *The functions $f(t)$, $f'(t)$ are a.p.;*

(3) *There exists a solution, $z(t)$, bounded and u.w.c.*

Then the minimal solution, $\tilde{x}(t)$, is w.a.p. and Y-a.p. Moreover, if $z(t)$, bounded, is u.c., then $\tilde{x}(t)$ is a.p.

VI (*Almost-periodicity theorem for the eigensolutions*). *Assume that:*

(1) *The operator $A(t)$ is periodic;*

(2) *The embedding of X in Y is compact.*

Then every bounded eigensolution, $u(t)$, is w.a.p. and Y-a.p. If $u(t)$ is bounded and u.c., then $u(t)$ is a.p.

OBSERVATION I. *If $A(t) = I$, then the hypothesis of (weak or strong) uniform continuity of the bounded solution $z(t)$, or $u(t)$, can be eliminated.*

OBSERVATION II. By (1.11), $z(t)$ *u.w.c.* $\Rightarrow \tilde{x}(t)$ *u.w.c.*

Moreover, setting

$$\varphi(x;\, \tau) = \sup_J \|x(t+\tau) - x(t)\|, \tag{1.15}$$

we have, by (1.11) and (1.12),

$$\varphi(x;\, \tau) \leq \varphi(z;\, \tau). \tag{1.16}$$

Hence, $z(t)$ *u.c.* $\Rightarrow \tilde{x}(t)$ *u.c.*

OBSERVATION III. In the example given above, hypotheses (1) and (2) of theorem V are satisfied if the functions $a_{jk}(t) = \{a_{jk}(t,\zeta);\, \zeta \in \Omega\}$, $a_0(t) = \{a_0(t,\zeta);\, \zeta \in \Omega\}$ and their derivatives $a'_{jk}(t)$, $a'_0(t)$ are $L^\infty(\Omega)$-a.p. and if the function $\Phi(t) = \{\Phi(t,\zeta,\xi);\, (\zeta,\xi) \in \Omega \times \Omega\}$ and its derivative $\Phi'(t)$ are $L^2(\Omega \times \Omega)$-a.p. Moreover, hypothesis (2) of theorem VI is satisfied if Ω is *bounded*.

2. FUNDAMENTAL EQUATIONS. INITIAL VALUE PROBLEM

(*a*) We shall now study equation (1.1) under more general assumptions than those made in § 1. More precisely, we shall assume that, in place of (i_1), (i_2), (i_3), the following conditions are satisfied:

(i'_1) $x(t) \in C^0(J;\, Y)$ *and is w.c.;*

(i'_2) $A(t) \in C^0(J;\, \mathscr{A})$ *and is self-adjoint;*

(i'_3) $f(t)$ *is w.c.*

Let us prove first the following *lemma*.

$$(1.1) \Leftrightarrow (2.1)\ i\frac{d}{dt}(x(t),v)_Y = (A(t)x(t) + f(t),v) \qquad (\forall v \in X). \tag{2.1}$$

Proof. (1.1) $\Rightarrow$ (2.1). Setting, in fact, $h(t) = \psi(t)v$, with $\psi(t) \in \mathscr{D}(J)$ (that is, $\in C^\infty(J)$ and with compact support), $v \in X$, we obtain from (1.1)

$$\int_J i(x(t),v)_Y \psi'(t)dt = -\int_J (A(t)x(t) + f(t),v)\psi(t)dt,$$

that is, (2.1), being $\psi(t)$ arbitrary.

(2.1) $\Rightarrow$ (1.1). We obtain, by (2.1), $\forall \sigma > 0$,

$$\begin{aligned} i\left(\frac{x(t+\sigma) - x(t)}{\sigma},\, v\right)_Y &= \frac{1}{\sigma}\int_t^{t+\sigma} (A(\eta)x(\eta) + f(\eta),\, v)d\eta \\ &= \left(\frac{1}{\sigma}\int_t^{t+\sigma} \{A(\eta)x(\eta) + f(\eta)\}d\eta,\, v\right) \end{aligned} \tag{2.2}$$

and also

$$i\int_J\left(\frac{x(t+\sigma)-x(t)}{\sigma},\, h(t)\right)_Y dt = -i\int_J\left(x(t),\, \frac{h(t)-h(t-\sigma)}{\sigma}\right)_Y dt$$
$$= \int_J\left(\frac{1}{\sigma}\int_t^{t+\sigma}\{A(\eta)x(\eta)+f(\eta)\}d\eta,\, h(t)\right)dt.$$

Relation (1.1) follows letting $\sigma \to 0$.

I. *Let $x(t)$, $z(t)$ be two solutions corresponding, respectively, to the known terms $f(t)$ and $g(t)$. The following first fundamental equation then holds on any bounded interval $a \vdash\!\dashv b$:*

$$i[(x(t),z(t))_Y]_a^b = \int_a^b\{(f(t),z(t))-(x(t),g(t))\}dt. \tag{2.3}$$

Let us set, for $\sigma > 0$, $\rho > 0$,

$$\begin{aligned} x_\sigma(t) &= \frac{1}{\sigma}\int_t^{t+\sigma} x(\eta)d\eta, \qquad z_\rho(t) = \frac{1}{\rho}\int_t^{t+\rho} z(\eta)d\eta,\\ f_\sigma(t) &= \frac{1}{\sigma}\int_t^{t+\sigma} f(\eta)d\eta + \frac{1}{\sigma}\int_t^{t+\sigma}(A(\eta)-A(t))x(\eta)d\eta,\\ g_\rho(t) &= \frac{1}{\rho}\int_t^{t+\rho} g(\eta)d\eta + \frac{1}{\rho}\int_t^{t+\rho}(A(\eta)-A(t))z(\eta)d\eta. \end{aligned} \tag{2.4}$$

Hence, in X and $\forall t \in J$,

$$\begin{aligned} x_\sigma(t) &\overset{*}{\to} x(t), \qquad z_\rho(t) \overset{*}{\to} z(t),\\ f_\sigma(t) &\overset{*}{\to} f(t), \qquad g_\rho(t) \overset{*}{\to} g(t). \end{aligned} \tag{2.5}$$

Moreover, the Y-derivatives $x'_\sigma(t)$, $z'_\rho(t)$ exist, are continuous, and we have

$$x'_\sigma(t) = \frac{x(t+\sigma)-x(t)}{\sigma}, \qquad z'_\rho(t) = \frac{z(t+\rho)-z(t)}{\rho}. \tag{2.6}$$

By (2.4) and (2.2) (and by the analogous relation for $z(t)$), it follows that

$$\begin{aligned} i(x'_\sigma(t),z_\rho(t))_Y &= (A(t)x_\sigma(t) + f_\sigma(t),\, z_\rho(t)),\\ -i(x_\sigma(t),z'_\rho(t))_Y &= (x_\sigma(t),\, A(t)z_\rho(t)+g_\rho(t)), \end{aligned}$$

that is,

$$i\frac{d}{dt}(x_\sigma(t),z_\rho(t))_Y = (f_\sigma(t),\, z_\rho(t))-(x_\sigma(t),g_\rho(t)).$$

Hence,

$$i[(x_\sigma(t),z_\rho(t))_Y]_a^b = \int_a^b\{(f_\sigma(t),z_\rho(t))-(x_\sigma(t),g_\rho(t))\}dt \tag{2.7}$$

and, letting $\sigma \to 0$, by (2.4) and (2.5),

$$i[(x(t),z_\rho(t))_Y]_a^b = \int_a^b \{(f(t),z_\rho(t)) - (x(t),g_\rho(t))\}dt.$$

Now letting $\rho \to 0$, we obtain (2.3).

Corollary I. *The eigensolutions, $u(t)$, of the homogeneous equation have constant Y-norm.*

From (2.3) (with $f(t)=g(t)=0$, $x(t)=z(t)=u(t)$) it follows, in fact, setting $a=0$, $b=t$, that

$$\|u(t)\|_Y = \|u(0)\|_Y. \tag{2.8}$$

Corollary II. *Under the assumptions* (i$'_1$), (i$'_2$), (i$'_3$), (i$_4$), *the uniqueness theorem for the initial value problem holds.*

In fact, if $x(t)$, $z(t)$ are two solutions of (1.1) such that

$$x(0) = z(0) = x_0 \in X,$$

then $u(t)=z(t)-x(t)$ is a solution of (1.3) with $u(0)=0$. It follows from (2.8) that $u(t)=0$.

(*b*) Let us now solve the initial value problem.

II. *Assume (besides (i'_1)) that $A(t)$, $f(t)$ satisfy conditions* (i$_2$), (i$_3$).

The initial value problem, $x(0)=x_0$, admits then ($\forall x_0 \in X$) one and only one solution, $x(t)$, which, moreover, is continuous on J.

The uniqueness of the solution has already been proved. To prove its existence, we shall use the Faedo-Galerkin method.

Let $\{y_n\}$, $y_n \in X$, be a sequence of linearly independent vectors, complete in X and in Y. We may assume that $(y_j,y_k)_Y=\delta_{jk}$.

Setting

$$x_n(t) = \sum_1^n {}_k\, \alpha_{nk}(t)y_k \tag{2.9}$$

we consider the "approximating system"

$$i(x'_n(t),y_j)_Y = (A(t)x_n(t)+f(t),y_j) \qquad (j = 1,\cdots, n), \tag{2.10}$$

that is,

$$i\alpha'_{nj}(t) = \sum_1^n {}_k\, \alpha_{nk}(t)(A(t)y_k,y_j)+(f(t),y_j),$$

with the initial condition

$$x_n(0) = w_n,$$

assuming that

$$w_n \to x_0, \qquad \|w_n\| \leq \|x_0\|. \tag{2.11}$$

Multiplying (2.10) by $\bar{\alpha}'_{nj}(t)$ and adding, we obtain

$$i\|x'_n(t)\|_Y^2 = (A(t)x_n(t)+f(t),x'_n(t)).$$

Hence,

$$-i\|x'_n(t)\|_Y^2 = (x'_n(t),\ A(t)x_n(t)+f(t))$$

and, consequently,

$$(A(t)x_n(t)+f(t),x'_n(t))+(x'_n(t),A(t)x_n(t)+f(t)) = 0.$$

We obtain, therefore,

$$\frac{d}{dt}(A(t)x_n(t),x_n(t))-(A'(t)x_n(t),x_n(t))+(f(t),x'_n(t))+(x'_n(t),f(t)) = 0,$$

that is,

$$\frac{d}{dt}\{(A(t)x_n(t),x_n(t))+2\mathscr{R}(f(t),x_n(t))\} = (A'(t)x_n(t),x_n(t))+2\mathscr{R}(f'(t),x_n(t)).$$

Hence, integrating on $0\vdash\!\dashv t$, we have

$$(2.12)\quad [(A(t)x_n(t),x_n(t))+2\mathscr{R}(f(t),x_n(t))]_0^t$$
$$= \int_0^t \{(A'(\eta)x_n(\eta),x_n(\eta))+2\mathscr{R}(f'(\eta),x_n(\eta))\}d\eta.$$

Let us consider the interval $0\vdash\!\dashv T$, $T>0$ (the same procedure holds for the interval $-T\vdash\!\dashv 0$). Setting

$$A_T = \max_{0\vdash\dashv T}\|A(t)\|,\qquad f_T = \max_{0\vdash\dashv T}\|f(t)\|,$$

we obtain, by (1.2),

$$\nu\|x_n(t)\|^2 \le A_T\|w_n\|^2+2f_T(\|x_n(t)\|+\|w_n\|)$$
$$+\int_0^t\{\|A'(\eta)\|\,\|x_n(\eta)\|^2+2\|f'(\eta)\|\,\|x_n(\eta)\|\}d\eta,$$

that is,

$$\nu\|x_n(t)\|^2-2f_T\|x_n(t)\|$$
$$-\left\{A_T\|w_n\|^2+2f_T\|w_n\|+\int_0^t\{\|A'(\eta)\|\,\|x_n(\eta)\|^2+2\|f'(\eta)\|\,\|x_n(\eta)\|\}d\eta\right\} \le 0.$$

Hence (since, by (1.2), $\nu\le A_T$),

$$(2.13)\quad \|x_n(t)\| \le \frac{2}{\nu}\left\{f_T^2+\nu\left[A_T\|w_n\|^2+2f_T\|w_n\|+\int_0^t\{\|A'(\eta)\|\,\|x_n(\eta)\|^2\right.\right.$$
$$\left.\left.+2\|f'(\eta)\|\,\|x_n(\eta)\|\}d\eta\right]\right\}^{1/2}$$
$$\le \frac{2}{\nu}\left\{(f_T+A_T\|w_n\|)^2+\nu\int_0^t\{\|A'(\eta)\|\,\|x_n(\eta)\|^2\right.$$
$$\left.+2\|f'(\eta)\|\,\|x_n(\eta)\|\}d\eta\right\}^{1/2}.$$

Having chosen $\varepsilon > 0$ arbitrarily, let $\psi_{n\varepsilon}(t)$ be a solution >0 of the equation

$$(2.14)\quad \psi_{n\varepsilon}^2(t) = \frac{4}{\nu^2}\left\{(f_T + A_T\|w_n\| + \varepsilon)^2 + \nu\int_0^t \{\|A'(\eta)\|\psi_{n\varepsilon}^2(\eta) + 2\|f'(\eta)\|\psi_{n\varepsilon}(\eta)\}d\eta\right\},$$

that is, of the equation

$$\psi_{n\varepsilon}'(t) = \frac{2}{\nu}\{\|A'(t)\|\psi_{n\varepsilon}(t) + 2\|f'(t)\|\}$$

with the initial condition

$$\psi_{n\varepsilon}(0) = \frac{2}{\nu}(f_T + A_T\|w_n\| + \varepsilon)$$

Hence,

$$(2.15)\quad \psi_{n\varepsilon}(t) = \frac{2}{\nu}(f_T + A_T\|w_n\| + \varepsilon)\exp\left[\frac{2}{\nu}\int_0^t \|A'(\eta)\|d\eta\right] + \frac{4}{\nu}\int_0^t \|f'(\eta)\| \exp\left[\frac{2}{\nu}\int_\eta^t \|A'(\theta)\|d\theta\right]d\eta.$$

We prove now that

$$(2.16)\quad \|x_n(t)\| < \psi_{n\varepsilon}(t).$$

Since, obviously, $\|x_n(0)\| < \psi_{n\varepsilon}(0)$, by the continuity $\|x_n(t)\| < \psi_{n\varepsilon}(t)$ on an interval $0 \vdash \delta_{n\varepsilon}$ (which we shall assume to be the greatest possible), with $\delta_{n\varepsilon} \le T$. Moreover, necessarily $\delta_{n\varepsilon} = T$, because, if $\delta_{n\varepsilon} < T$, then, by (2.13) and (2.14), $\|x_n(\delta_{n\varepsilon})\| < \psi_{n\varepsilon}(\delta_{n\varepsilon})$ and consequently $\|x_n(t)\| < \psi_{n\varepsilon}(t)$ in a right-hand neighborhood of $\delta_{n\varepsilon}$. Hence, $\forall\varepsilon > 0$, $\|x_n(t)\| < \psi_{n\varepsilon}(t)$ on the whole of $0 \vdash\!\dashv T$. Letting $\varepsilon \to 0$, we obtain, by (2.15) and (2.16), the inequalities

$$(2.17)\quad \begin{aligned} \|x_n(t)\| &\le \frac{2}{\nu}(f_T + A_T\|w_n\|)\exp\left[\frac{2}{\nu}\int_0^t \|A'(\eta)\|d\eta\right] \\ &\quad + \frac{4}{\nu}\int_0^t \|f'(\eta)\| \exp\left[\frac{2}{\nu}\int_\eta^t \|A'(\theta)\|d\theta\right]d\eta \\ &\le K_T\left\{\|w_n\| + \|f(0)\| + \int_0^T \|f'(\eta)\|d\eta\right\} \\ &\le K_T\left\{\|x_0\| + \|f(0)\| + \int_0^T \|f'(\eta)\|d\eta\right\} \\ &= B_T, \end{aligned}$$

K_T being a constant >0, independent of n, x_0, and $f(t)$.

Hence, by (2.17) ($\tilde{k}$ being the embedding constant of X in Y),

$$\|x_n(t)\|_Y \leq \tilde{k}B_T = C_T. \tag{2.18}$$

Let us now prove that the functions $x_n(t)$ are Y-equally weakly continuous on $0 \vdash\!\dashv T$ (i.e., that, $\forall y \in Y$, the functions $(x_n(t), y)_Y$ are equally continuous on $0 \vdash\!\dashv T$).

Having chosen, in fact, $\varepsilon > 0$, arbitrarily, let $w_\varepsilon = \sum_{(1)j}^{(q_\varepsilon)} \sigma_{j\varepsilon} y_j$ be a linear combination of the vectors y_j such that

$$\| y - w_\varepsilon \|_Y \leq \varepsilon. \tag{2.19}$$

Assuming $n \geq q_\varepsilon$, we have then, by (2.10), (2.17), (2.18), and (2.19),

$$\begin{aligned} |(x_n(t+\tau) - x_n(t), y)_Y| &\leq |(x_n(t+\tau) - x_n(t), y - w_\varepsilon)_Y| + |(x_n(t+\tau) - x_n(t), w_\varepsilon)_Y| \\ &\leq 2C_T\varepsilon + \left| \int_t^{t+\tau} (A(\eta)x_n(\eta) + f(\eta), w_\varepsilon) d\eta \right| \\ &\leq 2C_T\varepsilon + (A_T B_T + f_T)(\varepsilon + \|y\|)|\tau|. \end{aligned}$$

The thesis then follows, observing that the number of functions $(x_n(t), y)_Y$, with $n < q_\varepsilon$, is finite.

By the theorem of Ascoli-Arzelà, we may assume that the sequence $\{(x_n(t), y_j)\}_Y$ converges uniformly on $0 \vdash\!\dashv T$, $\forall y_j$. The sequence $(x_n(t), y)_Y$ then converges uniformly on $0 \vdash\!\dashv T$, $\forall y \in Y$.

Having chosen, in fact, $\varepsilon > 0$ arbitrarily, let us determine

$$z_\varepsilon = \sum_{1}^{r_\varepsilon}{}_j \, \gamma_{j\varepsilon} y_j$$

in such a way that

$$\| y - z_\varepsilon \|_Y \leq \varepsilon.$$

Then, by (2.18),

$$\begin{aligned} |(x_m(t), y)_Y - (x_n(t), y)_Y| &\leq |(x_m(t), z_\varepsilon)_Y - (x_n(t), z_\varepsilon)_Y| + |(x_m(t) - x_n(t), y - z_\varepsilon)_Y| \\ &\leq |(x_m(t), z_\varepsilon)_Y - (x_n(t), z_\varepsilon)_Y| + 2C_T\varepsilon, \end{aligned}$$

which proves the thesis.

Hence,

$$\lim_{n \to 0}{}^{*} x_n(t) \underset{Y}{=} x(t), \qquad \text{uniformly on } 0 \vdash\!\dashv T, \tag{2.20}$$

and the limit function $x(t)$ is Y-w.c.

Let us now prove that the functions $x_n(t)$ are also X-equally w.c. on $0 \vdash\!\dashv T$. We have, $\forall x \in X$ and $\forall y \in Y$,

$$|(x, y)_Y| \leq \|x\|_Y \|y\|_Y \leq \tilde{k}\|y\|_Y \|x\|. \tag{2.21}$$

Hence,

$$(x,y)_Y = (x,Gy), \tag{2.22}$$

where G is called Green's operator, from Y to X.

The set GY is dense in X. If it were not, then there would exist $v_0 \in X$, with $v_0 \neq 0$, such that $(v_0,Gy)=0$, $\forall y \in Y$; it follows that $(v_0,y)_Y=0$ and, consequently, $v_0=0$, which is absurd.

Having taken $v \in X$ and chosen $\varepsilon>0$ arbitrarily, let us now determine

$$v_\varepsilon = \sum_1^\infty{}_j\, \theta_{j\varepsilon} y_j \in Y,$$

in such a way that

$$\|v-Gv_\varepsilon\| \leq \varepsilon.$$

Moreover, by (2.17), we have

$$\begin{aligned}
&|(x_n(t+\tau),\, v)-(x_n(t),v)| \\
&\qquad \leq |(x_n(t+\tau),\, Gv_\varepsilon)-(x_n(t),Gv_\varepsilon)| + |(x_n(t+\tau)-x_n(t),\, v-Gv_\varepsilon)| \\
&\qquad \leq |(x_n(t+\tau),\, v_\varepsilon)_Y-(x_n(t),v_\varepsilon)_Y| + 2B_T\varepsilon.
\end{aligned}$$

Hence, from the equicontinuity of the functions $(x_n(t),\, v_\varepsilon)_Y$ follows the equicontinuity of the functions $(x_n(t),v)$.

By (2.17), we may therefore assume that the sequence $\{x_n(t)\}$ is weakly convergent, uniformly on $0 \vdash\!\dashv T$. Then, by (2.20),

$$\lim_{n\to\infty}{}^{*}\, x_n(t) = x(t), \qquad \text{uniformly on } 0 \vdash\!\dashv T, \tag{2.23}$$

and, consequently, $x(t)$ is w.c. By (2.11), we have $x(0)=x_0$.

Let us prove that $x(t)$ satisfies equation (1.1). By (2.17), moreover, (1.9) will be satisfied.

We have, in fact, $\forall y_j$, when $n \geq j$,

$$\begin{aligned}
i(x_n(t),y_j)_Y &= i(x_n(0),y_j)_Y + \left(\int_0^t \{A(\eta)x_n(\eta)+f(\eta)\}d\eta, y_j\right) \\
&= i(x_n(0),y_j)_Y + \int_0^t (A(\eta)x_n(\eta),y_j)d\eta + \int_0^t (f(\eta),y_j)d\eta.
\end{aligned}$$

Hence, letting $n \to \infty$, we obtain, by (2.11), (2.17), and (2.23),

$$\begin{aligned}
i(x(t),y_j)_Y &= i(x_0,y_j)_Y + \int_0^t (A(\eta)x(\eta),y_j)d\eta + \int_0^t (f(\eta),y_j)d\eta \\
&= i(x_0,y_j)_Y + \left(\int_0^t \{A(\eta)x(\eta)+f(\eta)\}d\eta, y_j\right).
\end{aligned}$$

Since the sequence $\{y_j\}$ is complete on X, it follows that, $\forall v \in X$,

$$i(x(t),v)_Y = i(x_0,v)_Y + \left(\int_0^t \{A(\eta)x(\eta)+f(\eta)\}d\eta,\, v\right) \tag{2.24}$$

that is, by (2.1), the thesis.

From (2.24), it follows also that

$$i(x(t+\tau)-x(t),\, x(t+\tau)-x(t))_Y = \left(\int_t^{t+\tau} \{A(\eta)x(\eta)+f(\eta)\}d\eta,\, x(t+\tau)-x(t)\right)$$

and, by (2.17),

$$\|x(t+\tau)-x(t)\|_Y^2 \leq 2B_T(A_T B_T+f_T)|\tau|,$$

that is, $x(t)$ is Y-continuous (and even Hölder-continuous, with exponent $\frac{1}{2}$).

Finally, let us prove that $x(t)$ is continuous. We shall begin by proving the continuity at the point $t=0$. Setting $z(t)=x(t)-x_0$, observe that we have, $\forall t \in 0 \longmapsto T$,

$$(2.25) \qquad x(t)-x_0 = \lim_{n\to\infty}{}^* \,(x_n(t)-w_n) = \lim_{n\to\infty}{}^* \, z_n(t),$$

$z_n(t)$ being the solution of the system

$$i(z_n'(t), y_j)_Y = (A(t)z_n(t)+(f(t)+A(t)w_n), y_j) \qquad (j = 1,\cdots, n)$$

with the initial condition

$$z_n(0) = 0.$$

It follows then from (2.12) that

$$\begin{aligned}(A(t)z_n(t), z_n(t)) = {} & -2\mathscr{R}(f(t)+A(t)w_n,\, z_n(t)) \\ & + \int_0^t \{(A'(\eta)z_n(\eta), z_n(\eta))+2\mathscr{R}(f'(\eta)+A'(\eta)w_n, z_n(\eta))\}d\eta\end{aligned}$$

that is, by (2.11) and (2.17),

$$(2.26) \quad \begin{aligned}\nu\|z_n(t)\|^2 \leq {} & 2|(f(t)+A(t)w_n,\, z_n(t))|+(B_T+\|x_0\|)^2 \int_0^t \|A'(\eta)\|d\eta \\ & +2(B_T+\|x_0\|)\int_0^t \{\|f'(\eta)\|+\|A'(\eta)\|\,\|x_0\|\}d\eta.\end{aligned}$$

Moreover, by (2.11), (2.23), and (2.25),

$$\lim_{n\to\infty} (f(t)+A(t)w_n,\, z_n(t)) = (f(t)+A(t)x_0,\, x(t)-x_0)$$

and, consequently, by (2.25) and (2.26),

$$\begin{aligned}\nu\|x(t)-x_0\|^2 \leq {} & 2|(f(t)+A(t)x_0,\, x(t)-x_0)|+(B_T+\|x_0\|)^2 \int_0^t \|A'(\eta)\|d\eta \\ & +2(B_T+\|x_0\|)\int_0^t \{\|f'(\eta)\|+\|A'(\eta)\|\,\|x_0\|\}d\eta.\end{aligned}$$

Hence

$$\lim_{t\to 0} \|x(t)-x_0\| = 0,$$

that is, $x(t)$ is continuous at the point $t=0$.

We now prove that $x(t)$ is continuous at any other point $t_0 \in 0 \vdash\!\dashv T$.

Let, in fact, $z(t)$ be the solution satisfying the initial condition

$$z(t_0) = x(t_0).$$

For what has been already proved, $z(t)$ is w.c. on $0 \vdash\!\dashv T$; moreover, it is continuous at t_0. By the uniqueness theorem for the solution of the initial value problem (corollary II to theorem I), we have, on the other hand, $z(t) = x(t)$ and the thesis is therefore proved.

III. *Under the assumptions of theorem II, the following second fundamental equation holds:*

$$(2.27) \quad [(A(t)x(t),z(t)) + (f(t),z(t)) + (x(t),g(t))]_a^b$$
$$= \int_a^b \{(A'(\eta)x(\eta),z(\eta)) + (f'(\eta),z(\eta)) + (x(\eta),g'(\eta))\} d\eta.$$

Relation (2.27) can be immediately proved if $x(t)$, $z(t)$ have continuous X-derivatives.

From (2.1) it follows in fact, in this case, that

$$i(x'(t),z'(t))_Y = (A(t)x(t) + f(t),z'(t)),$$
$$-i(x'(t),z'(t))_Y = (x'(t),A(t)z(t) + g(t)).$$

Hence,

$$(A(t)x(t) + f(t),z'(t)) + (x'(t),A(t)z(t) + g(t)) = 0,$$

that is,

$$\frac{d}{dt}\{(A(t)x(t),z(t)) + (f(t),z(t)) + (x(t),g(t))\}$$
$$= (A'(t)x(t),z(t)) + (f'(t),z(t)) + (x(t),g'(t)).$$

Relation (2.27) follows, integrating on $a \vdash\!\dashv b$.

Let us now consider the general case, introducing the functions $x_\sigma(t)$, $f_\sigma(t)$, $z_\rho(t)$, $g_\rho(t)$ defined by (2.4). These functions have continuous derivatives and we obtain

$$(2.28) \quad [(A(t)x_\sigma(t),z_\rho(t)) + (f_\sigma(t),z_\rho(t)) + (x_\sigma(t),g_\rho(t))]_a^b$$
$$= \int_a^b \{(A'(\eta)x_\sigma(\eta),z_\rho(\eta)) + (f'_\sigma(\eta),z_\rho(\eta)) + (x_\sigma(\eta),g'_\rho(\eta))\} d\eta.$$

Observe now that

$$(2.29) \quad f'_\sigma(t) = \frac{f(t+\sigma) - f(t)}{\sigma} + \frac{A(t+\sigma) - A(t)}{\sigma} x(t+\sigma)$$
$$- \frac{A'(t)}{\sigma} \int_t^{t+\sigma} x(\eta) d\eta \to f'(t).$$

Letting σ, $\rho \to 0$, from (2.28) and (2.29) follows then (2.27).

3. A MINIMAX THEOREM

Proof. Let us prove, first of all, the existence of the minimal solution.

Let $\{x_n(t)\}$ be a sequence minimizing the functional $\mu(x)$ in Γ_z; precisely

$$\mu(x_n) \downarrow \tilde{\mu}. \tag{3.1}$$

Since the sequence $\{x_n(0)\}$ is bounded, we assume, forthwith, that it converges weakly; hence,

$$\lim_{n\to\infty}{}^* x_n(0) = \tilde{x}(0),$$

that is, $\forall p$,

$$\lim_{n\to\infty}{}^* x_{p+n}(0) = \tilde{x}(0). \tag{3.2}$$

Observe now that, if $\rho_k \geq 0$, $\sum_{(1)k}^{(q)} \rho_k = 1$, the function

$$w(t) = \sum_{1}^{q}{}_k \rho_k x_{p+k}(t)$$

satisfies (1.1), $\in \Gamma_z$, and we have, by (3.1),

$$\mu(w) \leq \sum_{1}^{q}{}_k \rho_k \mu(x_{p+k}) \leq \mu(x_{p+1}).$$

Hence we can obtain, $\forall p$, by a theorem of Mazur, a solution

$$w_p(t) = \sum_{1}^{q_p}{}_k \rho_{pk} x_{p+k}(t) \qquad \left(\rho_{pk} \geq 0, \sum_{1}^{q_p}{}_k \rho_{pk} = 1\right)$$

such that

$$\|w_p(0) - \tilde{x}(0)\| \leq \frac{1}{p}, \qquad \mu(w_p) \downarrow \tilde{\mu}.$$

We may therefore assume that

$$\lim_{n\to\infty} x_n(0) = \tilde{x}(0). \tag{3.3}$$

Let $\tilde{x}(t)$ be the solution corresponding to the initial value $\tilde{x}(0)$.

By relation (1.9) (of continuous dependence on the data), we obtain then, uniformly on every bounded interval,

$$\lim_{n\to\infty} x_n(t) = \tilde{x}(t). \tag{3.4}$$

Hence, by (1.12), (1.13), and (3.1),

$$\varphi(\tilde{x}; v, \tau) \leq \varphi(z; v, \tau), \qquad \mu(\tilde{x}) \leq \tilde{\mu}.$$

Therefore $\tilde{x}(t) \in \Gamma_z$, and we have, necessarily, $\mu(\tilde{x}) = \tilde{\mu}$.

We now prove the uniqueness of the minimal solution.

Let $\tilde{x}(t)$ and $x_0(t)$ be two different solutions, $\in \Gamma_z$, such that

$$\mu(\tilde{x}) = \mu(x_0) = \tilde{\mu}. \tag{3.5}$$

Then $u(t) = \tilde{x}(t) - x_0(t)$ satisfies the homogeneous equation and we have, by the fundamental equation (a′) at § 1, since the embedding of X in Y is continuous,

$$\inf_J \|\tilde{x}(t) - x_0(t)\| \geq \frac{1}{\tilde{k}} \inf_J \|\tilde{x}(t) - x_0(t)\|_Y = \frac{1}{\tilde{k}} \|\tilde{x}(0) - x_0(0)\|_Y = \theta > 0 \tag{3.6}$$

(it cannot be $\theta = 0$, by the uniqueness of the solution of the initial value problem).

Hence, $\forall t \in J$,

$$\|\tilde{x}(t) - x_0(t)\| \geq \frac{\theta}{\tilde{\mu}} \max\{\|\tilde{x}(t)\|, \|x_0(t)\|\},$$

and (by (4.29) and (4.30), Chapter 3),

$$\left\|\frac{\tilde{x}(t) + x_0(t)}{2}\right\| \leq \sqrt{1 - \frac{\theta^2}{4\tilde{\mu}^2}} \max\{\|\tilde{x}(t)\|, \|x_0(t)\|\} \leq \sqrt{1 - \frac{\theta^2}{4\tilde{\mu}^2}}\, \tilde{\mu}$$

that is,

$$\mu\left(\frac{\tilde{x} + x_0}{2}\right) < \tilde{\mu},$$

which is absurd.

4. ALMOST-PERIODICITY THEOREM

Proof. (*a*) As, by assumption, $z(t)$ is w.u.c., the minimal solution, $\tilde{x}(t)$, is also w.u.c.

We shall now prove that $\tilde{x}(t)$ is w.a.p. extending the procedure given by Favard for linear ordinary a.p. systems.

Let $c = \{c_n\}$ be an arbitrary real sequence. We have to prove that it is possible to select from c a subsequence $s = \{s_n\}$ such that the sequence $\{\tilde{x}(t + s_n)\}$ is uniformly weakly convergent.

We assume, by the almost-periodicity, that

$$\begin{aligned} &\lim_{n\to\infty} A(t+s_n) = A_s(t), \qquad &&\lim_{n\to\infty} A'(t+s_n) = A'_s(t), \\ &\lim_{n\to\infty} f(t+s_n) = f_s(t), \qquad &&\lim_{n\to\infty} f'(t+s_n) = f'_s(t). \end{aligned} \tag{4.1}$$

uniformly.

As, moreover, $\tilde{x}(t)$ is bounded and w.u.c. and the space X is separable, we assume also, by the theorem of Ascoli-Arzelà, that the sequence $\{\tilde{x}(t+s_n)\}$ is weakly convergent, uniformly on every bounded interval. Hence, $\forall v \in X$,

$$\lim_{n\to\infty} (\tilde{x}(t+s_n),v) = (\tilde{x}_s(t),v) \tag{4.2}$$

uniformly on every bounded interval.

Since, $\forall y \in Y$, $(\tilde{x}(t),y)_Y=(x(t),Gy)$, the sequence $\{\tilde{x}(t+s_n)\}$ is also Y-weakly convergent, uniformly on every bounded interval. Moreover, $\forall n$,

$$\int_J \{i(\tilde{x}(t+s_n),\, h'(t))_Y+(\tilde{x}(t+s_n),\, A(t+s_n)h(t))+(f(t+s_n),\, h(t))\}dt = 0$$

and, letting $n\to\infty$, we have

$$\int_J \{i(\tilde{x}_s(t),h'(t))_Y+(\tilde{x}_s(t),A_s(t)h(t))+(f_s(t),h(t))\}dt = 0.$$

Hence $\tilde{x}_s(t)$ is a solution of the equation of Schrödinger type:

$$\int_J \{i(x(t),h'(t))_Y+(A_s(t)x(t)+f_s(t),\, h(t))\}dt = 0. \tag{4.3}$$

We have also, by (1.10) and (1.11),

$$\begin{gathered} \mu(\tilde{x}_s) \le \mu(\tilde{x}) = \tilde{\mu}, \\ \varphi(\tilde{x}_s;\, v,\tau) \le \varphi(\tilde{x};\, v,\tau) \qquad (\forall v \in X,\ \tau \in J). \end{gathered} \tag{4.4}$$

Let us prove that in (4.4) the $=$ sign holds and that $\tilde{x}_s(t)$ is the minimal solution of (4.3) in $\Gamma_{z,s}$, where $\Gamma_{z,s}$ is the set of the bounded solutions $x(t)$ of (4.3) such that $\varphi(x;\, v,\tau)\le\varphi(z;\, v,\tau)$ $(\forall v \in X,\ \tau \in J)$.

Consider, in fact, the sequence $-s=\{-s_n\}$; we may assume that

$$\lim_{n\to\infty}{}^{*}\, \tilde{x}_s(t-s_n) = w(t),$$

uniformly on every bounded interval, where $w(t)$ is a solution of (1.1) satisfying the inequalities

$$\begin{gathered} \mu(w) \le \mu(\tilde{x}_s) \le \mu(\tilde{x}), \\ \varphi(w;\, v,\tau) \le \varphi(\tilde{x}_s;\, v,\tau) \le \varphi(\tilde{x};\, v,\tau) \le \varphi(z;\, v,\tau). \end{gathered} \tag{4.5}$$

Hence $w \in \Gamma_z$ and, by the uniqueness of the minimal solution, we have $w(t)=\tilde{x}(t)$. It follows then, by (4.5), that

$$\mu(\tilde{x}_s) = \mu(\tilde{x}), \qquad \varphi(\tilde{x}_s;\, v,\tau) = \varphi(\tilde{x};\, v,\tau). \tag{4.6}$$

Let us prove that $\tilde{x}_s(t)$ is the minimal solution in $\Gamma_{z,s}$.

In fact, let $v_s(t)$ be such a minimal solution. If it were $v_s(t)\ne\tilde{x}_s(t)$, we would have $\mu(v_s)<\mu(\tilde{x}_s)$, which is absurd. We could in fact, prove, as before, that, in this case, there would exist a solution $\tilde{w}(t)$ of (1.1), with $\tilde{w} \in \Gamma_z$, $\mu(\tilde{w}) \le \mu(v_s) < \mu(\tilde{x}_s) = \mu(\tilde{x})$. Hence, $v_s(t) = \tilde{x}_s(t)$.

In order to prove that $\tilde{x}(t)$ is w.a.p., we must show that, $\forall v \in X$, relation (4.2) holds uniformly on J.

Assume that the convergence is not uniform for a certain $v_0 \in X$. There exist then a number $\rho > 0$ and three sequences $\{t_n\}$, $\{s_{n1}\} \subset \{s_n\}$, $\{s_{n2}\} \subset \{s_n\}$ such that

$$|(\tilde{x}(t_n + s_{n1}), v_0) - (\tilde{x}(t_n + s_{n2}), v_0)| \ge \rho. \tag{4.7}$$

We can assume that the sequences $\{\sigma_{n1}\} = \{t_n + s_{n1}\}$, $\{\sigma_{n2}\} = \{t_n + s_{n2}\}$ are regular, that is, that we have uniformly (for $j = 1, 2$)

$$\begin{aligned} \lim_{n \to \infty} A(t + \sigma_{nj}) &= A_{\sigma j}(t), & \lim_{n \to \infty} A'(t + \sigma_{nj}) &= A'_{\sigma j}(t), \\ \lim_{n \to \infty} f(t + \sigma_{nj}) &= f_{\sigma j}(t), & \lim_{n \to \infty} f'(t + \sigma_{nj}) &= f'_{\sigma j}(t). \end{aligned} \tag{4.8}$$

We may assume, moreover, that we have, uniformly on every bounded interval,

$$\lim_{n \to \infty}{}^{*} \tilde{x}(t + \sigma_{nj}) = \tilde{x}_{\sigma j}(t),$$

where $\tilde{x}_{\sigma j}(t)$ is the minimal solution, in $\Gamma_{z,\sigma j}$, of the equation

$$\int_J \{i(x(t), h'(t))_Y + (A_{\sigma j}(t)x(t) + f_{\sigma j}(t), h(t))\} dt = 0.$$

Now, we obtain

$$A_{\sigma 1}(t) = A_{\sigma 2}(t), \qquad f_{\sigma 1}(t) = f_{\sigma 2}(t). \tag{4.9}$$

In fact, by (4.8) and the uniform convergence,

$$\begin{aligned} \|A(t + \sigma_{n1}) - A(t + \sigma_{n2})\| &\le \|A(t + t_n + s_{n1}) - A(t + t_n + s_n)\| \\ &\quad + \|A(t + t_n + s_n) - A(t + t_n + s_{n2})\| \\ &\le \sup_J \|A(t + s_{n1}) - A(t + s_n)\| \\ &\quad + \sup_J \|A(t + s_n) - A(t + s_{n2})\| \to 0. \end{aligned}$$

The second of (4.9) is proved in the same way.

By the uniqueness of the minimal solution, $\tilde{x}_{\sigma 1}(t) = \tilde{x}_{\sigma 2}(t)$. Hence, when $t = 0$,

$$\lim_{n \to \infty} (\tilde{x}(\sigma_{n1}), v_0) = (\tilde{x}_{\sigma 1}(0), v_0) = (\tilde{x}_{\sigma 2}(0), v_0) = \lim_{n \to \infty} (\tilde{x}(\sigma_{n2}), v_0),$$

which contradicts (4.7).

The uniform convergence is therefore proved and $\tilde{x}(t)$ is w.a.p.

It is obvious that $\tilde{x}(t)$ is also Y-w.a.p., since

$$(\tilde{x}(t), y)_Y = (\tilde{x}(t), Gy), \qquad \forall y \in Y.$$

(*b*) Let us prove that $\tilde{x}(t)$ is Y-a.p. For this, we shall use the fundamental relation (a) of § 1. Since $f(t)$ is a.p. and $\tilde{x}(t)$ is w.a.p., the scalar product $(f(t),\tilde{x}(t))$ is a.p. It follows, by (a), that the derivative $(d/dt)\|\tilde{x}(t)\|_Y^2$ is a.p.

Hence, $\|\tilde{x}(t)\|_Y^2$, being bounded, is a.p.

Let us now consider an arbitrary sequence $\{s_n\}$ and the sequence $\{\tilde{x}(t+s_n)\}$, $\{A(t+s_n)\}$, $\{f(t+s_n)\}$. We assume, by the almost-periodicity, that

$$\lim_{n\to\infty}{}^{*}\ \tilde{x}(t+s_n) = \tilde{x}_s(t) \tag{4.10}$$

uniformly, and that (4.1) hold.

Hence $\tilde{x}_s(t)$ is, by (1.1), a w.a.p. solution of the equation

$$\int_J \{i(x(t),h'(t))_Y + (A_s(t)x(t)+f_s(t),h(t))\}dt = 0 \tag{4.11}$$

and also the norm $\|x_s(t)\|_Y$ is a.p.

By theorem XI of Chapter 3 it then follows that $\tilde{x}(t)$ is Y-a.p.

(*c*) Let us prove that, if the bounded solution $z(t)$ is u.c., $\tilde{x}(t)$ is a.p. It follows in fact from (2.1) that, $\forall v \in X$,

$$i\frac{d}{dt}(\tilde{x}(t),v)_Y = (A(t)\tilde{x}(t)+f(t),v),$$

that is,

$$i(\tilde{x}(t),v)_Y = i(\tilde{x}(0),v)_Y + \left(\int_0^t \{A(\eta)\tilde{x}(\eta)+f(\eta)\}d\eta, v\right). \tag{4.12}$$

Moreover,

$$(\tilde{x}(t),v)_Y = (G\tilde{x}(t),v)$$

and therefore, by (4.12),

$$iG\tilde{x}(t) = iG\tilde{x}(0) + \int_0^t \{A(\eta)x(\eta)+f(\eta)\}d\eta. \tag{4.13}$$

Since

$$\|G\tilde{x}(t+\tau)-G\tilde{x}(t)\| \le \|G\|\,\|\tilde{x}(t+\tau)-\tilde{x}(t)\|_Y,$$

$G\tilde{x}(t)$ is an a.p. function. Hence, the integral

$$\int_0^t \{A(\eta)\tilde{x}(\eta)+f(\eta)\}d\eta$$

is a.p.

By observation II, § 1, the minimal solution $\tilde{x}(t)$ is u.c. Hence $A(t)\tilde{x}(t)+f(t)$ is u.c. and, therefore, a.p.

Since $f(t)$ is a.p., it follows that $A(t)\tilde{x}(t)$ is a.p.

Let us now observe that, by (1.2) and the almost-periodicity of $A(t)$,

$$0 < \nu \leq \|A(t)\| \leq N < +\infty.$$

Hence there exists the inverse operator $A^{-1}(t)$ and

$$N^{-1} \leq \|A^{-1}(t)\| \leq \nu^{-1}.$$

Moreover,

$$\begin{aligned}\|A^{-1}(t+\tau)-A^{-1}(t)\| &= \|A^{-1}(t+\tau)(A(t)-A(t+\tau))A^{-1}(t)\| \\ &\leq \nu^{-2}\|A(t+\tau)-A(t)\|.\end{aligned}$$

Therefore, $A^{-1}(t)$ is $\mathscr{A}$-a.p. and the function $\tilde{x}(t)=A^{-1}(t)A(t)\tilde{x}(t)$ is a.p.

(*d*) *Assume*, finally, *that* $A(t)=I$. In this case we can prove that, *if* $f(t)$ *and* $f'(t)$ *are a.p. and if there exists a bounded solution*, $z(t)$, *then the minimal solution*, $\tilde{x}(t)$, *is a.p.*

Let $c=\{c_n\}$ be an arbitrary real sequence. We prove first that there exists a subsequence $s=\{s_n\}$ such that the sequence $\{x(t+s_n)\}$ is weakly convergent, uniformly on every bounded interval $-T\leq t\leq T$. For this we shall use a procedure given by Zaidman [2] for a similar question concerning wave equation.

Let us observe that every solution $x(t)$ can be put, by the fundamental relation (b′), in the form

$$x(t) = U(t)x(0)+w(t) \tag{4.14}$$

where $U(t)x(0)$ satisfies the homogeneous equation with initial value $x(0)$, and $w(t)$ satisfies (1.1) with the initial condition $w(0)=0$; $U(t)$ is an X-strongly continuous unitary operator.

Now let $s=\{s_n\}$ be a subsequence of $c=\{c_n\}$ such that

$$\lim_{n\to\infty}{}^{*}\, \tilde{x}(s_n) = \tilde{x}_s(0) \tag{4.15}$$

and that relations (4.10) hold.

Since $\{\tilde{x}(t+s_n)\}$ satisfies the equation with operator I and known term $\{f(t+s_n)\}$, we obtain, by (4.14),

$$\tilde{x}(t+s_n) = U(t)\tilde{x}(s_n)+w_n(t) \qquad (w_n(0) = 0). \tag{4.16}$$

Moreover, $\forall v \in X$,

$$(U(t)\tilde{x}(s_n),v) = (\tilde{x}(s_n),U^{-1}(t)v).$$

Hence, by (4.15) and because of the continuity of $U^{-1}(t)v$, the sequence $\{U(t)\tilde{x}(s_n)\}$ is weakly convergent, uniformly on every bounded interval $-T\leq t\leq T$.

Moreover, by (1.9) and (4.16), on the same interval,

$$\|w_n(t)-w_m(t)\| \leq M_T\left\{\|f(s_n)-f(s_m)\|+\int_{-T}^{T}\|f'(\eta+s_n)-f'(\eta+s_m)\|d\eta\right\},$$

i.e., the sequence $\{w_n(t)\}$ converges uniformly on $-T \vdash\!\dashv T$. Hence the sequence $\{\tilde{x}(t+s_n)\}$ is weakly convergent, uniformly on every bounded interval.

In order to prove that $\tilde{x}(t)$ is w.a.p., we must now show that such a convergence is uniform on J and this can be done following the same procedure given at (a).

Finally, let us prove that $\tilde{x}(t)$ is a.p.

This will be done by the same procedure followed for the wave equation (Chapter 5, § 4 (b_2)) and for the integration of a.p. functions (Chapter 4, § 3). Assume in fact, that there exist a sequence $\{s_n\}$ and a number $\rho>0$ such that

$$\|\tilde{x}(s_j)-\tilde{x}(s_k)\| \geq \rho \qquad (j \neq k). \tag{4.17}$$

We assume that $\{s_n\}$ is regular with respect to $f(t)$, $f'(t)$ and that

$$\lim{}^* \tilde{x}(t+s_n) = \tilde{x}_s(t),$$

where $\tilde{x}_s(t)$ is the minimal solution corresponding to $f_s(t)=\lim_{n\to\infty} f(t+s_n)$.

Let us set

$$\tilde{x}(t+s_j)-\tilde{x}(t+s_k) = u_{jk}(t)+z_{jk}(t)$$

where $u_{jk}(t)$ satisfies the homogeneous equation, with the initial condition $u_{jk}(0)=\tilde{x}(s_j)-\tilde{x}(s_k)$ and $z_{jk}(t)$ is the solution corresponding to the known term $f(t+s_j)-f(t+s_k)$ and to the initial value $z_{jk}(0)=0$.

Then, by (b′), (1.9), (2.17), and (4.17), $\forall t \in J$,

$$\|\tilde{x}(t+s_j)-\tilde{x}(t+s_k)\| \geq \|u_{jk}(t)\|-\|z_{jk}(t)\| =$$

$$\|u_{jk}(0)\|-\|z_{jk}(t)\| \geq \rho-N_t\left\{\|f(s_j)-f(s_k)\|+\left|\int_0^t\|f'(\eta+s_j)-f'(\eta+s_k)\|d\eta\right|\right\},$$

with $0<N_t<+\infty$, independent of j, k.

It follows that

$$\max \lim_{(j,k)\to\infty} \|\tilde{x}(t+s_j)-\tilde{x}(t+s_k)\| \geq \rho.$$

By theorem XII of Chapter 3, $\tilde{x}(t)$ is therefore a.p.

5. ALMOST-PERIODICITY THEOREM OF THE EIGENSOLUTIONS

Proof. Assume that $A(t)$ is periodic with period 1 and let $u(t)$ be an eigensolution of (1.3); then $u(t+n)-u(t+m)$, $\forall$ integers n and m, satisfies the same equation.

Hence, by (a′),

$$\|u(t+n)-u(t+m)\|_Y = \|u(n)-u(m)\|_Y. \tag{5.1}$$

Let us now assume that $u(t)$ is bounded. Since the sequence $\{u(n)\}$ is bounded and the embedding of X in Y is compact, $\{u(n)\}$ is Y-r.c.

Setting now, in the space $K=C^0(J;Y) \cap L^\infty(J;Y), \tilde{u}(s)=\{u(t+s);\, t\in J\}$, we obtain

$$\|\tilde{u}(n)-\tilde{u}(m)\|_K = \sup_J \|u(t+n)-u(t+m)\|_Y = \|u(n)-u(m)\|_Y.$$

The sequence $\{\tilde{u}(n)\}$ is therefore r.c. and (by property (ζ), Chapter 1, § 2) we conclude that $u(t)$ is Y-a.p.

Let us now prove that $u(t)$ is w.a.p.

By (2.22) and since GY is dense in X, we can, $\forall\, v \in X$ and $\varepsilon>0$, find $z_\varepsilon \in Y$ such that $\|v-Gz_\varepsilon\| \leq \varepsilon$.

Setting $M=\sup_J \|u(t)\|$, we obtain

$$\begin{aligned}|(u(t+\tau)-u(t),v)| &\leq |(u(t+\tau)-u(t),Gz_\varepsilon)|+2\varepsilon M\\ &= |(u(t+\tau)-u(t),z_\varepsilon)_Y|+2\varepsilon M,\end{aligned}$$

that is, $(u(t),v)$ is a.p., $\forall v \in X$.

Finally, let us assume that $u(t)$, bounded, is u.c.

We can then prove that $u(t)$ is a.p. by the same procedure given at (c), § 4, for the proof of the almost-periodicity of $\tilde{x}(t)$.

Assume finally that $A(t)=I$ and that the embedding of X in Y is compact: then every eigensolution $u(t)$ is a.p.

In fact, by (b′) of § 1, $\|u(t)\|=\|u(0)\|$, that is, $\|u(t)\|$ is a.p. Being $u(t)$ w.a.p., then we deduce that it is also a.p.

CHAPTER 7

THE WAVE EQUATION WITH NONLINEAR DISSIPATIVE TERM

1. INTRODUCTION AND STATEMENTS

(*a*) In this chapter we shall study the *wave equation with dissipative term*:

$$(1.1)\quad \sum_{j,k}^{1\cdots m} \frac{\partial}{\partial \zeta_j}\left(a_{jk}(\zeta)\frac{\partial x(t,\zeta)}{\partial \zeta_k}\right) - a_0(\zeta)x(t,\zeta) - \frac{\partial^2 x(t,\zeta)}{\partial t^2} + f(t,\zeta) = \beta\left(\frac{\partial x(t,\zeta)}{\partial t}\right),$$

where, as usual, $t \in J$, $\zeta = \{\zeta_1, \cdots, \zeta_m\} \in \Omega$, open, bounded, and connected set of R^m.

All functions are supposed to be *real*. We shall assume that the coefficients $a_{jk}(\zeta)$, $a_0(\zeta)$ satisfy the same conditions set, in Chapter 5, for the wave equation, and use here the same notations. We assume also, for the time being, that the known term $f(t) = \{f(t,\zeta);\ \zeta \in \Omega\}$ satisfies the condition:

$$(1.2)\qquad f(t) \in L^1_{\text{loc}}(J; L^2).$$

Partial differential equation (1.1) is a particular case of an abstract equation considered by Lions and Strauss [1]. The theorem proved by Lions and Strauss, when applied to (1.1), guarantees the existence and uniqueness of the solution of the initial-boundary value problem under the hypothesis that $\beta(\eta)$ is a nondecreasing function of class $C^0(J)$, with the same asymptotic behavior as $\eta|\eta|^\sigma$, $\sigma \geq 0$. The statements and the proofs corresponding to this behavior will be given, respectively, in part (*b*) of this section and in § 5.

Regarding the dissipative term, *we shall now only assume that* $\beta(\eta)$ *is a nondecreasing function of the variable* $\eta \in \overline{a\,b}$, *with* $-\infty \leq a < 0 < b \leq +\infty$, *such that* $\beta(0^-) \leq 0 \leq \beta(0^+)$ *and*

$$(1.3)\qquad \begin{aligned} &\lim_{\eta \to a^+} \beta(\eta) = -\infty \qquad \textit{if } a > -\infty, \\ &\lim_{\eta \to b^-} \beta(\eta) = +\infty \qquad \textit{if } b < +\infty. \end{aligned}$$

The function $\beta(\eta)$ *can therefore have discontinuities of the first kind in a sequence* $\{\eta_s\}$ *of points.*

The results obtained, under these assumptions, by Amerio and Prouse [1], [2], are the object of the present part (*a*) and of §§ 2, 3, 4.

We give, first, a *suitable definition of solution*, suggested by the presence of discontinuities of $\beta(\eta)$. Precisely, considering an interval $0 \leq t \leq T$ and the cylinder $Q = 0 \vdash\!\dashv T \times \Omega$, we shall say that $x(t,\zeta)$ is a *solution* on Q if:

(i_1) $x(t,\zeta) \in H^1(Q)$ (i.e., if $x, x_t, x_{\zeta_k} \in L^2(Q)$);

(i_2) *The distributions*

$$\frac{\partial^2 x}{\partial t^2},\ A(\zeta)x(t,\zeta) = \sum_{j,k}^{1\cdots m} \frac{\partial}{\partial \zeta_j}\left(a_{jk}(\zeta)\,\frac{\partial x(t,\zeta)}{\partial \zeta_k}\right) - a_0(\zeta)x(t,\zeta)$$

are functions $\in L^2(Q)$;

(i_3) *We have, a.e. on* Q,

$$a < x_t(t,\zeta) < b, \tag{1.4}$$

$$A(\zeta)x(t,\zeta) - x_{tt}(t,\zeta) + f(t,\zeta) \in \beta((x_t(t,\zeta))^-) \vdash\!\dashv \beta((x_t(t,\zeta))^+). \tag{1.5}$$

Relation (1.5) means that, at the point $(\bar{t},\bar{\zeta})$, if $x_t(\bar{t},\bar{\zeta}) = \bar{\eta}$ and if $\beta(\eta)$ is continuous at $\bar{\eta}$, then in $(\bar{t},\bar{\zeta})$ equation (1.1) holds; if instead $\bar{\eta} = \eta_s$, then

$$\beta(\eta_s^-) \leq A(\bar{\zeta})x(\bar{t},\bar{\zeta}) - x_{tt}(\bar{t},\bar{\zeta}) + f(\bar{t},\bar{\zeta}) \leq \beta(\eta_s^+).$$

Observe that (1.5) may be written in the form (1.1), provided it is understood that *at the point* η_s *the function* $\beta(\eta)$ *may take on any value of the interval* $\beta(\eta_s^-) \vdash\!\dashv \beta(\eta_s^+)$. In what follows, we shall *always* assume this convention.

Let us pose, for equation (1.1), the *initial-boundary* value problem: find a solution $x = x(t,\zeta)$ satisfying the *initial conditions*

$$x(0,\zeta) = x_0(\zeta), \qquad x_t(0,\zeta) = y_0(\zeta) \qquad (\zeta \in \Omega) \tag{1.6}$$

and the *boundary condition*

$$x(t,\zeta)\big|_{\zeta \in \partial\Omega} = 0 \qquad (0 \leq t \leq T). \tag{1.7}$$

This problem corresponds to the motion of a vibrating membrane, with *fixed edge*, under the action of an *external force* $f(t,\zeta)$ and of a *resistance* which is a *nondecreasing* (and eventually also *discontinuous*) *function of the velocity*; moreover, *if b is finite, the velocity cannot exceed the value b* (corresponding to an *infinite resistance*); *the same occurs for a*. The problem here treated is obviously of interest also in the *theory of controls*.

Setting $x(t) = \{x(t,\zeta);\ \zeta \in \Omega\}$, $x'(t) = \{x_t(t,\zeta);\ \zeta \in \Omega\}$, $x''(t) = \{x_{tt}(t,\zeta);\ \zeta \in \Omega\}$, $Ax(t) = \{A(\zeta)x(t,\zeta);\ \zeta \in \Omega\}$, and $\beta(x'(t)) = \{\beta(x_t(t,\zeta));\ \zeta \in \Omega\}$, equation (1.1) can also be written in the *operational* form:

$$Ax(t) - x''(t) + f(t) = \beta(x'(t)). \tag{1.8}$$

Let us denote by Γ the class of functions $z(t) = \{z(t,\zeta);\ \zeta \in \Omega\}$ such that:

(ii$_1$) $z(t) \in C^0(0 \vdash\!\dashv T; E)$;
(ii$_2$) $z'(t) \in L^2(0 \vdash\!\dashv T; E)$;
(ii$_3$) $Az(t) \in L^2(0 \vdash\!\dashv T; L^2)$.

Hence $z(t) \in \Gamma$ means that $z(t)$, $z'(t)$ are continuous from $0 \vdash\!\dashv T$ to H_0^1 and L^2, respectively, that $z'(t)$, $z''(t)$, $Az(t)$ are square-integrable from $0 \vdash\!\dashv T$, respectively, to H_0^1, L^2, L^2.

Consider now *only* the solutions $x(t)$ which belong to Γ. Then the *initial-boundary* value problem defined above corresponds to the following *initial* value problem for equation (1.8): *given x_0 and y_0, find a solution $x(t) \in \Gamma$, satisfying the initial conditions*

$$x(0) = x_0, \qquad x'(0) = y_0. \tag{1.9}$$

It is obvious that the solutions obtained in this class are *more regular* than those obtained in Chapter 5; the reason for the choice of the class Γ will become evident during the proofs of theorems I, II, III.

Setting

$$\bar\beta(\eta) = \begin{cases} 0 & \text{when } \eta = 0 \\ \beta(\eta^-) & \text{when } \eta > 0 \\ \beta(\eta^+) & \text{when } \eta < 0, \end{cases} \tag{1.10}$$

we shall prove (at § 2) the following *existence and uniqueness theorem.*

I. *Assume that:*

(1) *$\beta(\eta)$ is nondecreasing on $a \overline{\quad} b$ and satisfies conditions* (1.3)*;*
(2) *$x_0 \in H_0^1$, $Ax_0 \in L^2$;*
(3) *$y_0 \in H_0^1$, $a < y_0(\zeta) < b$ (a.e. on Ω), $\bar\beta(y_0) \in L^2$;*
(4) *$f(0) \in L^2$, $f'(t) \in L^1(0 \vdash\!\dashv T;\ L^2)$.*

Then the initial value problem (1.8) *and* (1.9) *admits, in the functional class Γ, one and only one solution.*

It follows, by (1.1), that

$$\beta(x_t(t,\zeta)) \in L^2(Q). \tag{1.11}$$

Observe that conditions (i$_1$), (i$_2$), (i$_3$) and (ii$_1$), (ii$_2$), (ii$_3$) can obviously be referred to any interval $T_1 \vdash\!\dashv T_2$ and to any cylinder $T_1 \vdash\!\dashv T_2 \times \Omega$.

Hence, by theorem I, if $t_0 \in J$ is arbitrarily fixed, *the solution corresponding to the initial values*

$$x(t_0) = x_0, \qquad x'(t_0) = y_0$$

can be prolonged on the whole interval $t_0 \vdash\!- +\infty$.

We shall say, in accordance with the definitions given above, that a solution $x(t)$ is *bounded if it is defined on all J and if* (cf. (7.24), Chapter 4):

$$(1.12)\quad \sup_J \{\|Ax(t)\|^2_{L^2(L^2)} + \|x'(t)\|^2_{L^2(E)}\}^{1/2}$$

$$= \sup_J \left\{\int_0^1 (\|Ax(t+\eta)\|^2_{L^2} + \|x'(t+\eta)\|^2_E)d\eta\right\}^{1/2} = M_x < +\infty.$$

One proves that *$x(t)$ bounded $\Rightarrow$ $x(t)$ E-u.c. and $\mathscr{R}_{x(t)}$ E-r.c.*

We state now two theorems (which will be proved at §§ 3, 4) concerning *bounded* or *a.p.* solutions of (1.8). Recall that $f(t)$ *$L^2(L^2)$-bounded* means:

$$(1.13)\qquad \sup_J \|f(t)\|_{L^2(L^2)} = \sup_J \left\{\int_0^1 \|f(t+\eta)\|^2_{L^2} d\eta\right\}^{1/2} < +\infty.$$

II (*Uniqueness theorem of the bounded solution*). *Assume that:*

(1) *$\beta(\eta)$ is strictly increasing on $a \frown b$, satisfies conditions* (1.3), *and is continuous at the point $\eta = 0$;*

(2) *$f(t)$ is $L^2(L^2)$-bounded.*

Then there exists at most one solution, $x(t)$, which is bounded on J.

III (*Almost-periodicity theorem*). *Assume that $f(t)$ is L^2-S^2 w.a.p. and that the hypotheses of theorem II are verified.*

Then, if $x(t)$ is a bounded solution, $x(t)$ is E-a.p., whereas $x'(t)$ and $Ax(t)$ are, respectively, E-S^2 w.a.p. and L^2-S^2 w.a.p.

Corollary. *If*, in addition to the assumptions made in theorem III, *$f(t)$ is periodic with period Θ, then the bounded solution $x(t)$* (if it exists) *is periodic with period Θ.*

(*b*) The almost-periodicity theorem III extends to the nonlinear wave equation (1.1) the first theorem of Favard (cf. Chapter 6, § 1, (*a*)), generalized by Amerio [1] to a.p. nonlinear ordinary systems: if $f(t)$ is a.p., the almost-periodicity of a bounded solution is, in fact, a consequence of its uniqueness. No existence theorem of a bounded solution is however given.

As mentioned at (*a*), § 5 is concerned with the study of equation (1.1) under much more strict conditions on the nonlinear term $\beta(x_t)$. We shall, however, develop the theory in the *energy space*, give *existence* and *uniqueness* theorems of a bounded solution, and analyze the asymptotic behavior, as $t \to +\infty$, of the solutions; then, if $f(t)$ is a.p., we shall obtain an *existence* and *uniqueness* theorem of a.p. solutions.

Assume, from now on, that *$\beta(\eta)$ is a continuous nondecreasing function on J, with $\beta(0) = 0$, and that there exist $p \geq 2$ and $c_1 > 0$, $c_2 > 0$, $\bar{\eta} > 0$ such that*

$$(1.14)\qquad c_1|\eta|^p \leq \eta\beta(\eta) \leq c_2|\eta|^p, \qquad when\ |\eta| \geq \bar{\eta}.$$

Assume, moreover, that *the known term* $f(t)$, $t \in J$, *satisfies the condition*

$$(1.15)\qquad f(t) \in L^r_{\text{loc}}(J; L^r), \qquad \textit{with } r = \frac{p}{p-1},$$

that is,

$$\|f(t)\|_{L^r(L^r)} = \left\{\int_0^1 \|f(t+\eta)\|^r_{L^r} d\eta\right\}^{1/r} < +\infty, \qquad \forall t \in J.$$

We assume, furthermore, that Ω *satisfies the cone property* (Sobolev [1]; cf. Lions [2]); this occurs, for example, if $\partial\Omega$ can be represented locally by means of functions satisfying Lipschitz conditions—in particular, if $\partial\Omega$ is constituted by a finite number of smooth $(m-1)$-dimensional surfaces.

An interesting case in which the hypotheses given above are satisfied corresponds to the motion of a vibrating membrane subject to the external force $f(t,\zeta)$ and to a *resistance* $\beta(x_t)$ *of viscous type for small values of the velocity and of hydraulic type for large values*:

$$(1.16)\qquad \beta(x_t) = \gamma_1 x_t + \gamma_2 |x_t| x_t \qquad (\gamma_1, \gamma_2 > 0).$$

Following the definition given by Lions and Strauss, we shall say that $x(t)$ is a *solution* on $0 \vdash\!\dashv T$ of the equation

$$(1.17)\qquad Ax(t) - x''(t) + f(t) = \beta(x'(t))$$

if

(i$'_1$) $x(t) \in L^\infty(0 \vdash\!\dashv T; E)$, $x'(t) \in L^p(0 \vdash\!\dashv T; L^p)$;

(i$'_2$) $x(t)$ *satisfies the equation*

$$(1.18)\qquad \int_0^T \{-(x'(t)), h'(t))_{L^2} + (x(t), h(t))_{H^1_0} + \langle \beta(x'(t)), h(t)\rangle - \langle f(t), h(t)\rangle\} dt = 0,$$

$\forall$ *test function* $h(t) \in L^1(0 \vdash\!\dashv T; E) \cap L^p(0 \vdash\!\dashv T; L^p)$, *with compact support on* $0 \frown T$.

The symbol $\langle \cdot, \cdot \rangle$ denotes the duality between the spaces L^p and L^r, that is,

$$\langle u, v\rangle = \int_\Omega u(\zeta) v(\zeta) d\zeta \qquad (\forall u \in L^r,\ v \in L^p).$$

The above definition is obviously very close to that given in Chapter 5 for the wave equation; contrary to what was assumed at (*a*), $x''(t)$ and $Ax(t)$ will, in general, be distributions and not functions.

Let us now state the existence and uniqueness theorem given by Lions and Strauss for the initial value problem.

IV. *Under the conditions defined by* (1.14) *and* (1.15) *for* $\beta(\eta)$ *and* $f(t)$, *equation* (1.18) *admits, on* $0 \vdash\!\dashv T$, $\forall T > 0$, *one and only one solution satisfying the arbitrary initial conditions*

$$(1.19) \qquad x(0) = x_0 \in H_0^1, \qquad x'(0) = y_0 \in L^2.$$

It follows, by (1.15), that the *initial value problem can be solved on any interval* $t_0 \vdash +\infty$, with arbitrary initial data given at $t = t_0$.

We observe that, by a theorem of Strauss [1], *any solution* $x(t)$ *of* (1.17) *is E-continuous and satisfies the following "energy" equation,* $\forall t_1, t_2 \in 0 \vdash\!\dashv T$:

$$(1.20) \quad \|x(t_2)\|_E^2 - \|x(t_1)\|_E^2 + 2\int_{t_1}^{t_2} \langle \beta(x'(t)), x'(t)\rangle dt = 2\int_{t_1}^{t_2} \langle f(t), x'(t)\rangle dt.$$

The analysis can be now developed by the following theorems, stated by Prouse [8] (in a functional space very close to the energy space).

We shall assume, in what follows, that

$$(1.21) \qquad 2 \le m \le 5, \qquad 2 \le p \le 2 + \frac{4}{m-1};$$

(hence, by Sobolev's embedding theorem: $x(t) \in L^2(0 \vdash\!\dashv T; E) \Rightarrow x(t) \in L^p(0 \vdash\!\dashv T; L^p))$.

We also assume that, in addition to (1.14), $\beta(\eta)$ *satisfies,* $\forall \eta_1, \eta_2 \in J$, *the conditions*

$$(1.22) \qquad c_3|\eta_2 - \eta_1|^{p-2} \le \frac{\beta(\eta_2) - \beta(\eta_1)}{\eta_2 - \eta_1} \le c_4(1 + |\eta_1|^{p-2} + |\eta_2|^{p-2}) \qquad (c_3, c_4 > 0).$$

Observe that, if $\beta(\eta)$ is defined by (1.16), then (1.22) and the second of (1.21) hold, provided that $m \le 5$.

We shall say, finally, that a solution $x(t)$ is *bounded* if it is *defined on all J and if*

$$(1.23) \qquad \sup_J \|x(t)\|_E < +\infty, \qquad \sup_J \|x'(t)\|_{L^p(L^p)} < +\infty,$$

that is, if $x(t)$ is E-bounded and $x'(t)$ is $L^p(L^p)$-bounded.

Let us state now the above mentioned results, where $J_0 = 0 \vdash +\infty$.

V. $x(t)$ *solution on* J_0 *and* $f(t)$ $L^r(L^r)$*-bounded on* $J_0 \Rightarrow x(t)$ *E-bounded on* J_0 *and* $x'(t)$ $L^p(L^p)$*-bounded on* J_0. *Precisely, if we set*

$$\sup_J \|x(t)\|_E = M_1 < +\infty, \qquad \sup_J \|x'(t)\|_{L^p(L^p)} = M_2 < +\infty,$$

the constants M_1, M_2 *depend only on* f, β, Ω, $\|x_0\|_{H_0^1}$, $\|y_0\|_{L^2}$.

VI. $x(t)$ *solution on* J_0 *and* $f(t)$ $L^r(L^r)$*-bounded and* $L^r(L^r)$*-u.c. on* $J_0 \Rightarrow x(t)$ *E-u.c. on* J_0.

VII (*Existence theorem*). *$f(t)$ $L^r(L^r)$-bounded and $L^r(L^r)$-u.c. $\Rightarrow \exists$ a solution $x(t)$, bounded and E-u.c.*

VIII. *If*

$$\max \lim_{t \to -\infty} \|f(t)\|_{L^r(L^r)} < +\infty, \tag{1.24}$$

there exists at most one solution $x(t)$, such that

$$\max \lim_{t \to -\infty} \|x(t)\|_E < +\infty. \tag{1.25}$$

Hence, *if $f(t)$ is $L^r(L^r)$-bounded, the uniqueness theorem of the bounded solutions holds.*

Observe that this result does not follow from theorem II, since the solutions are considered here in a wider functional class.

IX. *Assume that*

$$\max \lim_{t \to +\infty} \|f(t)\|_{L^r(L^r)} < +\infty. \tag{1.26}$$

Then, denoting by $x_1(t)$, $x_2(t)$, $t \in J_0$, any two solutions, we obtain

$$\lim_{t \to +\infty} \|x_2(t) - x_1(t)\|_E = 0. \tag{1.27}$$

X (*Almost-periodicity theorem*). *Assume that $f(t)$ is L^r-S^r a.p. Then there exists one and only one bounded solution, $x(t)$. Moreover, $x(t)$ and $x'(t)$ are, respectively, E-a.p. and L^p-S^p a.p. Finally, all the solutions E-converge asymptotically to $x(t)$, when $t \to +\infty$.*

OBSERVATION I. An existence theorem of a *periodic* solution of equation (1.17) (with $f(t)$ *periodic*) has been given by Prodi [5] under the same hypotheses made by Lions and Strauss, but with $\beta(\eta)$ strictly increasing. The functional class considered by Prodi for the solutions is slightly different from the one introduced above.

OBSERVATION II. With regard to the equation $x_{tt} - x_{\zeta\zeta} = f(t,\zeta,x)$, we recall the analysis made by Günzler [1], [2], [4] concerning, in particular, the *traces* on $t=0$ of those solutions which are a.p. in t (following the concept of *generalized a.p. function*, in the sense of Delsarte, Levitan, and Marčenko).

2. EXISTENCE AND UNIQUENESS THEOREM FOR THE INITIAL-BOUNDARY VALUE PROBLEM

(*a*) Let us begin by proving the uniqueness theorem.

Let $x(t)$, $y(t)$ be two solutions $\in \Gamma$ such that

$$x(0) = y(0) = x_0, \qquad x'(0) = y'(0) = y_0. \tag{2.1}$$

Then

$$(2.2)\qquad \begin{aligned} Ax(t)-x''(t)+f(t) &= \beta(x'(t)),\\ Ay(t)-y''(t)+f(t) &= \beta(y'(t)) \end{aligned}$$

and, if we set $w(t)=y(t)-x(t)$,

$$(2.3)\qquad Aw(t)-w''(t) = \beta(x'(t)+w'(t))-\beta(x'(t)),$$

where, by (2.1), $w(t)$ satisfies the initial conditions

$$(2.4)\qquad w(0) = 0, \qquad w'(0) = 0.$$

Consider the scalar product, in L^2, of (2.3) by $w'(t)$ and integrate between 0 and $t \in 0\vdash\!\dashv T$. By (2.4), and since, by (2.2),

$$\beta(x'(t)),\beta(y'(t)) \in L^2(0\vdash\!\dashv T;\ L^2),$$

we obtain

$$(2.5)\qquad \begin{aligned} \|w(t)\|_E^2 &= -2\int_0^t (\beta(x'(\eta))+w'(\eta))-\beta(x'(\eta)),w'(\eta))_{L^2}d\eta \\ &= -2\int_0^t d\eta\int_\Omega (\beta(x_t(\eta,\zeta)+w_t(\eta,\zeta))-\beta(x_t(\eta,\zeta)))w_t(\eta,\zeta)d\zeta. \end{aligned}$$

Observe now that we have, a.e. on Q,

$$(2.6)\qquad (\beta(x_t(t,\zeta)+w_t(t,\zeta))-\beta(x_t(t,\zeta)))w_t(t,\zeta) \geq 0.$$

This obviously occurs at a point $(\bar{t},\bar{\zeta})$ if it is $w_t(\bar{t},\bar{\zeta})=0$. Assume now, for instance, that $w_t(\bar{t},\bar{\zeta})>0$; setting then

$$\eta_1 = x_t(\bar{t},\bar{\zeta}), \qquad \eta_2 = x_t(\bar{t},\bar{\zeta})+w_t(\bar{t},\bar{\zeta}),$$

we obtain (since $\eta_1<\eta_2$)

$$\beta(\eta_1) \leq \beta(\eta_1^+) \leq \beta(\eta_2^-) \leq \beta(\eta_2).$$

Hence (2.6) is proved, $\forall w_t(t,\zeta)$.

By (2.5), $\forall t \in 0\vdash\!\dashv T$, it follows that

$$\|w(t)\|_E = 0$$

and the uniqueness theorem is proved.

(*b*) We now recall some results that will be utilized in the proof of the existence theorem to be given in section (*c*).

Let $0\leq\varphi(t) \in L^\infty(0\vdash\!\dashv T)$, $0\leq\omega(t)\in L^1(0\vdash\!\dashv T)$ *and, moreover,*

$$\varphi^2(t) \leq K^2+2\int_0^t \omega(\eta)\varphi(\eta)d\eta.$$

Then

$$(2.7)\qquad \varphi^2(t) \leq 2\left\{K^2+\left(\int_0^t \omega(\eta)d\eta\right)^2\right\}.$$

Set, $\forall \varepsilon > 0$,

$$\psi_\varepsilon^2(t) = (K+\varepsilon)^2 + 2\int_0^t \omega(\eta)\varphi(\eta)d\eta \qquad (\psi_\varepsilon(t) > 0).$$

It follows that $\varphi(t) < \psi_\varepsilon(t)$ and

$$\psi_\varepsilon(t)\psi_\varepsilon'(t) = \omega(t)\varphi(t) \le \omega(t)\psi_\varepsilon(t) \quad \text{(a.e.)},$$

$$\psi_\varepsilon(t) \le (K+\varepsilon) + \int_0^t \omega(\eta)d\eta.$$

Since ε is arbitrary, we have then

$$0 \le \varphi(t) \le K + \int_0^t \omega(\eta)d\eta \Rightarrow (2.7).$$

Let us now denote by $\{g_j\}$ the sequence of the eigensolutions of the operator A:

$$g_j \in H_0^1,\ (g_j,g_k)_{L^2} = \left(\frac{g_j}{\sqrt{\lambda_j}}, \frac{g_k}{\sqrt{\lambda_k}}\right)_{H_0^1} = \delta_{jk},$$

and by $\{\lambda_j\}$ $(\lambda_{j+1} \ge \lambda_j > 0, \lim_{j\to\infty} \lambda_j = +\infty)$ the sequence of the corresponding eigenvalues.

If $x \in L^2$, we have

$$(2.8) \qquad x = \sum_1^\infty{}_n (x,g_n)_{L^2} g_n, \qquad \sum_1^\infty{}_n (x,g_n)_{L^2}^2 = \|x\|_{L^2}^2 < +\infty.$$

Characteristic condition for x to belong to H_0^1 is that

$$(2.9) \qquad \begin{gathered} x = \sum_1^\infty{}_n \left(x, \frac{g_n}{\sqrt{\lambda_n}}\right)_{H_0^1} \frac{g_n}{\sqrt{\lambda_n}} = \sum_1^\infty{}_n \sqrt{\lambda_n}(x,g_n)_{L^2} \frac{g_n}{\sqrt{\lambda_n}}, \\ \|x\|_{H_0^1}^2 = \sum_1^\infty{}_n \lambda_n (x,g_n)_{L^2}^2 < +\infty. \end{gathered}$$

Let, moreover, $Ax \in L^2$: $\forall \varphi \in \mathscr{D}(\Omega)$, we have then $(Ax,\varphi)_{L^2} = \langle Ax,\varphi\rangle = -(x,\varphi)_{H_0^1}$; hence, $\forall g \in H_0^1$, $(Ax,g)_{L^2} = -(x,g)_{H_0^1}$.

Characteristic condition for $Ax \in L^2$ and $x \in H_0^1$ is that

$$(2.10) \quad Ax = \sum_1^\infty{}_n (Ax,g_n)_{L^2} g_n = -\sum_1^\infty{}_n (x,g_n)_{H_0^1} g_n = -\sum_1^\infty{}_n \lambda_n (x,g_n)_{L^2} g_n,$$

that is

$$(2.11) \qquad \|Ax\|_{L^2}^2 = \sum_1^\infty{}_n \lambda_n^2 (x,g_n)_{L^2}^2 < +\infty.$$

Let now, for a sequence $\{x_k\}$, $x_k \in H_0^1$ and $\|Ax_k\| \le M$; *the sequence itself is then r.c. in the norm of* H_0^1.

(c) Let us now prove the existence of a solution $x(t) \in \Gamma$ of the initial value problem; as $x(t)$ takes its values in the same functional spaces to which the initial values x_0 and y_0 belong, the theorem will obviously remain valid on the whole interval $0 \vdash\!\!\!- +\infty$.

The theorem will be proved, at first, under supplementary conditions on the function $\beta(\eta)$; more precisely, *we shall assume that* $\beta(\eta)$, $\beta'(\eta)$ *are continuous and bounded on* $a \overline{} b$, *with* $\beta'(\eta) \geq 0$.

The general case will be treated afterward.

For the proof, we shall use the Faedo-Galerkin method; in our case it is particularly useful to take as a "basis" the sequence $\{g_j\}$ of eigenfunctions of the operator A.

Setting

$$x_n(t) = \sum_1^n {}_j\, \alpha_{nj}(t) g_j \qquad (j = 1, \cdots, n), \tag{2.12}$$

we consider the system of "approximating equations" deducted from (1.8):

$$(x_n''(t), g_j)_{L^2} - (Ax_n(t), g_j)_{L^2} - (f(t), g_j)_{L^2} = -(\beta(x_n'(t)), g_j)_{L^2}, \tag{2.13}$$

with the initial conditions

$$x_n(0) = \sum_1^n {}_j\, (x_0, g_j)_{L^2} g_j \qquad ((x_0, g_j)_{L^2} = \alpha_{nj}(0)), \tag{2.14}$$

$$x_n'(0) = \sum_1^n {}_j\, (y_0, g_j)_{L^2} g_j \qquad ((y_0, g_j)_{L^2} = \alpha_{nj}'(0)). \tag{2.15}$$

Hence,

$$Ax_n(0) = -\sum_1^n {}_j\, \lambda_j (x_0, g_j)_{L^2} g_j. \tag{2.16}$$

From (2.14), (2.15), and (2.16), it follows that

$$\begin{aligned} &\|x_n(0)\|_{H_0^1} \leq \|x_0\|_{H_0^1}, && \|x_n'(0)\|_{L^2} \leq \|y_0\|_{L^2}, \\ &\|x_n'(0)\|_{H_0^1} \leq \|y_0\|_{H_0^1}, && \|Ax_n(0)\|_{L^2} \leq \|Ax_0\|_{L^2}. \end{aligned} \tag{2.17}$$

Multiplying (2.13) by $\alpha_{nj}'(t)$ and adding, we obtain

$$(x_n''(t), x_n'(t))_{L^2} - (Ax_n(t), x_n'(t))_{L^2} - (f(t), x_n'(t))_{L^2} = -(\beta(x_n'(t)), x_n'(t))_{L^2},$$

that is (since $-(Ax_n(t), x_n'(t))_{L^2} = (x_n(t), x_n'(t))_{H_0^1}$ and $(\beta(x_n'(t)), x_n'(t))_{L^2} \geq 0$),

$$\frac{1}{2}\frac{d}{dt}\|x_n(t)\|_E^2 \leq (f(t), x_n'(t))_{L^2}. \tag{2.18}$$

Hence,

$$\|x_n(t)\|_E^2 \leq \|x_n(0)\|_E^2 + 2\int_0^t \|f(\eta)\|_{L^2} \|x_n(\eta)\|_E \, d\eta, \tag{2.19}$$

from which, by (2.17) and (2.7),

$$\|x_n(t)\|_E^2 \le 2\left\{\|x_n(0)\|_E^2 + \left(\int_0^T \|f(\eta)\|_{L^2} d\eta\right)^2\right\} \tag{2.20}$$
$$\le 2\left\{\|x_0\|_{H_0^1}^2 + \|y_0\|_{L^2}^2 + \left(\int_0^T \|f(\eta)\|_{L^2} d\eta\right)^2\right\} = K_1^2,$$

where $K_1 \ge 0$ does not depend on $\beta(\eta)$ and n; $x_n(t)$ exists in $0 \vdash\!\!\!\dashv t$.

Let us multiply (2.13) by $-\lambda_j \alpha'_{nj}(t)$ and add the corresponding equations. Observing that $Ax'_n(t) = -\sum_{(1)j}^{(n)} \lambda_j \alpha'_{nj}(t) g_j$, we obtain

$$(x''_n(t), Ax'_n(t))_{L^2} - (Ax_n(t), Ax'_n(t))_{L^2} - (f(t), Ax'_n(t))_{L^2} \tag{2.21}$$
$$= -(\beta(x'_n(t)), Ax'_n(t))_{L^2},$$

that is,

$$\frac{d}{dt}\{\|x'_n(t)\|_{H_0^1}^2 + \|Ax_n(t)\|_{L^2}^2\} = -2(f(t), Ax'_n(t))_{L^2} + 2(\beta(x'_n(t)), Ax'_n(t))_{L^2}. \tag{2.22}$$

Observe now that, by the hypotheses on $\beta(\eta)$, $g \in H_0^1 \Rightarrow \beta(g) \in H_0^1$; hence,

$$(\beta(x'_n(t)), Ax'_n(t))_{L^2} = -\int_\Omega \left\{\beta'\left(\frac{\partial x_n(t,\zeta)}{\partial t}\right) \sum_{j,k}^{1\ldots m} a_{jk}(\zeta) \frac{\partial^2 x_n(t,\zeta)}{\partial t \partial \zeta_j} \frac{\partial^2 x_n(t,\zeta)}{\partial t \partial \zeta_k}\right. \tag{2.23}$$
$$\left. + \beta\left(\frac{\partial x_n(t,\zeta)}{\partial t}\right) a_0(\zeta) \frac{\partial x_n(t,\zeta)}{\partial t}\right\} d\zeta \le 0.$$

Setting

$$\psi_n(t) = \{\|x'_n(t)\|_{H_0^1}^2 + \|Ax_n(t)\|_{L^2}^2\}^{1/2}, \tag{2.24}$$

we obtain

$$\psi_n^2(t) \le \psi_n^2(0) - 2\int_0^t (f(\eta), Ax'_n(\eta))_{L^2} d\eta \tag{2.25}$$
$$= \psi_n^2(0) - 2(f(t), Ax_n(t))_{L^2} + 2(f(0), Ax_n(0))_{L^2}$$
$$+ 2\int_0^t (f'(\eta), Ax_n(\eta))_{L^2} d\eta$$
$$\le \|y_0\|_{H_0^1}^2 + \|Ax_0\|_{L^2}^2 + 2\|f(0)\|_{L^2}\|Ax_0\|_{L^2}$$
$$+ 2\left(\|f(0)\|_{L^2} + \int_0^T \|f'(\eta)\|_{L^2} d\eta\right)\psi_n(t) + 2\int_0^t \|f'(\eta)\|_{L^2}\psi_n(\eta) d\eta$$
$$= M_1 + 2M_2\psi_n(t) + 2\int_0^t \|f'(\eta)\|_{L^2}\psi_n(\eta) d\eta,$$

where M_1, M_2 do not depend on $\beta(\eta)$ and on n. Hence

$$(\psi_n(t)-M_2)^2 \leq M_1+M_2^2+2M_2\int_0^T \|f'(\eta)\|_{L^2}d\eta$$
$$+2\int_0^t \|f'(\eta)\|_{L^2}|\psi_n(\eta)-M_2|d\eta,$$

that is, by (2.24),

$$(2.26) \qquad \|x_n'(t)\|_{H_0^1} \leq K_2, \qquad \|Ax_n(t)\|_{L^2} \leq K_2,$$

where $K_2 \geq 0$ is independent of $\beta(\eta)$ and of n.

Let us now differentiate (2.13):

$$(2.27) \quad (x_n'''(t),g_j)_{L^2}-(Ax_n'(t),g_j)_{L^2}-(f'(t),g_j)_{L^2} = -(\beta'(x_n'(t))x_n''(t),g_j)_{L^2}.$$

Multiplying (2.27) by $a_{nj}''(t)$ and adding, we obtain

$$(2.28) \quad (x_n'''(t),x_n''(t))_{L^2}+(x_n''(t),x_n'(t))_{H_0^1} = (f'(t),x_n''(t))_{L^2}-(\beta'(x_n'(t))x_n''(t),x_n''(t))_{L^2},$$

that is,

$$(2.29) \quad \frac{d}{dt}\{\|x_n''(t)\|_{L^2}^2+\|x_n'(t)\|_{H_0^1}^2\} = 2(f'(t),x_n''(t))_{L^2}-2(\beta'(x_n'(t))x_n''(t)),x_n''(t))_{L^2}.$$

Observing that $(\beta'(x_n')x_n'',x_n'')_{L^2} \geq 0$, it follows, from (2.7), that

$$(2.30) \quad \|x_n''(t)\|_{L^2}^2+\|x_n'(t)\|_{H_0^1}^2 \leq 2\left\{\|x_n''(0)\|_{L^2}^2+\|x_n'(0)\|_{H_0^1}^2+\left(\int_0^T \|f'(\eta)\|_{L^2}d\eta\right)^2\right\}$$
$$\leq 2\left\{\|x_n''(0)\|_{L^2}^2+\|y_0\|_{H_0^1}^2+\left(\int_0^T \|f'(\eta)\|_{L^2}d\eta\right)^2\right\}.$$

On the other hand,

$$(2.31) \qquad (x_n''(0),g_j)_{L^2} = (Ax_n(0),g_j)_{L^2}+(f(0),g_j)_{L^2}-(\beta(x_n'(0)),g_j)_{L^2}$$

that is,

$$(2.32) \qquad x_n''(0) = Ax_n(0)+\sum_{1}^{n}{}_k (f(0),g_k)_{L^2}g_k-\sum_{1}^{n}{}_k (\beta(x_n'(0)),g_k)_{L^2}g_k.$$

Bearing in mind (2.15) and the fact that $0 \leq \beta'(\eta) \leq K_\beta < +\infty$, we have

$$(2.33) \quad \int_\Omega \left(\beta\left(\frac{\partial x_n(0,\zeta)}{\partial t}\right)-\beta(y_0(\zeta))\right)^2 d\zeta \leq K_\beta^2 \int_\Omega \left(\frac{\partial x_n(0,\zeta)}{\partial t}-y_0(\zeta)\right)^2 d\zeta$$
$$= K_\beta^2 \sum_{n+1}^{\infty}{}_k (y_0,g_k)_{L^2}^2.$$

Hence,

$$(2.34) \qquad \|\beta(x_n'(0))\|_{L^2} \leq \|\beta(y_0)\|_{L^2}+K_\beta\|y_0-x_n'(0)\|_{L^2}.$$

From (2.30), (2.32), and (2.34), we obtain

$$\begin{aligned}(2.35)\quad &\|x_n''(t)\|_{L^2}^2+\|x_n'(t)\|_{H_0^1}^2\\ &\le 2\left\{\|y_0\|_{H_0^1}^2+\left(\int_0^T\|f'(\eta)\|_{L^2}d\eta\right)^2+4\|Ax_0\|_{L^2}^2+4\|f(0)\|_{L^2}^2\right.\\ &\qquad\left.+4\|\beta(y_0)\|_{L^2}^2+4K_\beta^2\|y_0-x_n'(0)\|_{L^2}^2\right\}\\ &= 8\{\|\beta(y_0)\|_{L^2}^2+K_\beta^2\|y_0-x_n'(0)\|_{L^2}^2\}+K_3^2,\end{aligned}$$

where

$$K_3 = 2^{1/2}\left\{\|y_0\|_{H_0^1}^2+\left(\int_0^T\|f'(\eta)\|_{L^2}d\eta\right)^2+4\|Ax_0\|_{L^2}^2+4\|f(0)\|_{L^2}^2\right\}^{1/2}$$

is independent of $\beta(\eta)$ and n.

Denoting by K_1, K_2, K_3 quantities that do not depend on $\beta(\eta)$ and n, the following evaluations therefore hold, $\forall t \in 0 \mapsto T$:

$$\begin{aligned}(2.36)\qquad &\|x_n(t)\|_E \le K_1, \qquad \|x_n'(t)\|_{H_0^1} \le K_2,\\ &\|Ax_n(t)\|_{L^2} \le K_2,\\ &\|x_n''(t)\|_{L^2} \le 3\{\|\beta(y_0)\|_{L^2}+K_\beta\|y_0-x_n'(0)\|_{L^2}\}+K_3.\end{aligned}$$

As the sequence $\{Ax_n(t)\}$ is L^2-bounded, it follows that $\{x_n(t)\}$ is H_0^1-r.c. $\forall t \in 0 \mapsto T$; since, moreover, $\|x_n'(t)\|_{H_0^1} \le K_2$, the functions $x_n(t)$ are H_0^1-equicontinuous. Hence, by the (vectorial) theorem of Ascoli-Arzelà, we can assume that

$$(2.37)\qquad \lim_{n\to\infty} x_n(t) \underset{H_0^1}{=} x(t) \qquad \text{uniformly on } 0 \mapsto T.$$

In the same way it follows, from the second and fourth of (2.36) (the embedding of H_0^1 in L^2 being compact) that

$$(2.38)\qquad \lim_{n\to\infty} x_n'(t) \underset{L^2}{=} x'(t) \qquad \text{uniformly on } 0 \mapsto T.$$

The limit function $x(t)$ is therefore, by (2.37) and (2.38), E-continuous on $0 \mapsto T$.

By (2.36), it is possible to assume that

$$\begin{aligned}(2.39)\qquad &\lim_{n\to\infty}{}^{*}\, x_n'(t) \underset{L^2(0\mapsto T;H_0^1)}{=} x'(t),\\ &\lim_{n\to\infty}{}^{*}\, x_n''(t) \underset{L^2(0\mapsto T;L^2)}{=} x''(t),\\ &\lim_{n\to\infty}{}^{*}\, Ax_n(t) \underset{L^2(0\mapsto T;L^2)}{=} Ax(t).\end{aligned}$$

We recall now that, if a sequence $\{z_n(t)\}$ is such that $z_n(t) \in L^2(0 \mapsto T;X)$ (X Hilbert space), $\|z_n(t)\|_X \le K$, and

$$\lim_{n\to\infty}{}^{*}\, z_n(t) \underset{L^2(0\mapsto T;X)}{=} z(t),$$

then $\|z(t)\|_X \le K$ a.e.

The limit function $x(t)$ therefore satisfies, by (2.36) and (2.39), the inequalities (a.e.)

$$\begin{aligned} &\|x(t)\|_{H_0^1} \le K_1, \qquad \|x'(t)_{H_0^1}\| \le K_2, \\ &\|Ax(t)\|_{L^2} \le K_2, \qquad \|x''(t)\|_{L^2} \le 3\|\beta(y_0)\|_{L^2}+K_3, \end{aligned} \tag{2.40}$$

where K_1, K_2, K_3 do not depend on $\beta(\eta)$. Observe that $\|x''(t)\|_{L^2}$ has been estimated independently of K_β, whereas this did not occur for $\|x_n''(t)\|_{L^2}$.

As $x(t) \in C^0(0\vdash\!\dashv T;E)$, it follows from (2.39) that $x(t) \in \Gamma$.

By the Fischer-Riesz theorem, we can assume, by (2.38), that we have, a.e. on Q,

$$\lim_{n\to\infty} \frac{\partial x_n(t,\zeta)}{\partial t} = \frac{\partial x(t,\zeta)}{\partial t} \tag{2.41}$$

and consequently ($\beta(\eta)$ being continuous),

$$\lim_{n\to\infty} \beta\left(\frac{\partial x_n(t,\zeta)}{\partial t}\right) = \beta\left(\frac{\partial x(t,\zeta)}{\partial t}\right). \tag{2.42}$$

As $\beta(\eta)$ is bounded, it follows that, $\forall\psi \in L^2(Q)$,

$$\lim_{n\to\infty} \int_Q \beta\left(\frac{\partial x_n(t,\zeta)}{\partial t}\right)\psi(t,\zeta)dtd\zeta = \int_Q \beta\left(\frac{\partial x(t,\zeta)}{\partial t}\right)\psi(t,\zeta)dtd\zeta. \tag{2.43}$$

In particular, if $\psi(t,\zeta)=g_j(\zeta)\varphi(t)$ ($\varphi(t) \in C^0(0\vdash\!\dashv T)$),

$$\lim_{n\to\infty} \int_0^T (\beta(x_n'(t)),g_j)_{L^2}\varphi(t)dt = \int_0^T (\beta(x'(t)),g_j)_{L^2}\varphi(t)dt. \tag{2.44}$$

From (2.37), (2.38), (2.39), (2.44), it follows that

$$\begin{aligned} \lim_{n\to\infty} \int_0^T &\{(x_n''(t),g_j)_{L^2} - (Ax_n(t),g_j)_{L^2} + (\beta(x_n'(t)),g_j)_{L^2} - (f(t),g_j)_{L^2}\}\varphi(t)dt \\ &= \int_0^T \{(x''(t),g_j)_{L^2} - (Ax(t),g_j)_{L^2} + (\beta(x'(t)),g_j)_{L^2} - (f(t),g_j)_{L^2}\}\varphi(t)dt \end{aligned} \tag{2.45}$$

and, by (2.13),

$$x''(t) - Ax(t) + \beta(x'(t)) - f(t) = 0. \tag{2.46}$$

The function $x(t) \in \Gamma$ therefore satisfies the given equation (1.8). As it verifies obviously the initial conditions (1.9), the theorem is proved.

We observe, finally, that, by (2.40) and (2.46),

$$\|\beta(x'(t))\|_{L^2} \le 3\|\beta(y_0)\|_{L^2} + K_4, \tag{2.47}$$

K_4 being independent of $\beta(\eta)$.

Let us now prove the existence theorem in the general case.

First, we define a suitable sequence $\{\beta_n(\eta)\}$ of functions continuous and bounded on J, together with their first derivatives, with $\beta_n'(\eta) \ge 0$.

Let $\{\rho_n\}$, $\{\sigma_n\}$ be two sequences such that $0<\rho_n<\rho_{n+1}\to b$, $0>\sigma_n>\sigma_{n+1}\to a$; assume, moreover, that $\beta(\eta)$ is continuous in ρ_n and σ_n, $\forall n$.

Bearing in mind that $\beta(0^+)\geq 0$, $\beta(0^-)\leq 0$, we set, for $n=1, 2,\cdots$,

$$(2.48)\qquad \beta_n^+(\eta) = \begin{cases}\beta(\eta) & \text{when } 0<\eta<\rho_n\\ \beta(\rho_n) & \text{when } \rho_n\leq\eta<+\infty\\ 0 & \text{when } \eta\leq 0\end{cases}$$

$$\beta_n^-(\eta) = \begin{cases}0 & \text{when } \eta\geq 0\\ \beta(\eta) & \text{when } 0>\eta>\sigma_n\\ \beta(\sigma_n) & \text{when } -\infty<\eta\leq\sigma_n.\end{cases}$$

Let $\varphi(\eta)$ be a function $\in C^\infty(J)$ such that

$$(2.49)\qquad \varphi(\eta)\geq 0,\qquad \int_J \varphi(\eta)d\eta = 1,\qquad \operatorname{supp}\varphi(\eta)\subset \overline{0\,\theta},$$

where $\theta=\min(\rho_1, -\sigma_1)$.

Setting

$$(2.50)\qquad \tilde\beta_n^+(\eta) = n\int_J \beta_n^+(\eta-s)\varphi(ns)ds = n\int_0^{\theta/n}\beta_n^+(\eta-s)\varphi(ns)ds$$
$$= n\int_J \beta_n^+(s)\varphi(n(\eta-s))ds,$$
$$\tilde\beta_n^-(\eta) = n\int_J \beta_n^-(\eta+s)\varphi(ns)ds = n\int_0^{\theta/n}\beta_n^-(\eta+s)\varphi(ns)ds$$
$$= n\int_J \beta_n^-(s)\varphi(n(s-\eta))ds,$$

we have $\tilde\beta_n^+(\eta)$, $\tilde\beta_n^-(\eta)\in C^\infty(J)$. Moreover, $\tilde\beta_n^+(\eta)$, $\tilde\beta_n^-(\eta)$ are nondecreasing, and

$$(2.51)\qquad \tilde\beta_n^+(\eta)=0 \quad\text{when } \eta\leq 0,\qquad \tilde\beta_n^-(\eta)=0\quad\text{when } \eta\geq 0.$$

Setting

$$(2.52)\qquad \beta_n(\eta) = \tilde\beta_n^+(\eta)+\tilde\beta_n^-(\eta),$$

we have then, by (1.10),

$$(2.53)\qquad \beta_n(0)=0,\qquad 0\leq\beta_n(\eta)\leq\beta(\eta^-)=\bar\beta(\eta)\qquad\text{when } b>\eta\geq 0,$$
$$0\geq\beta_n(\eta)\geq\beta(\eta^+)=\bar\beta(\eta)\qquad\text{when } a<\eta\leq 0.$$

Observe that $\beta_n(\eta)$ is a nondecreasing function, bounded on J together with all its derivatives. Moreover, the following property holds.

Let us fix $\eta\in\overline{a\,b}$. $\forall\varepsilon>0$, there exist δ_ε and n_ε (depending also on η) such that, when $|\xi-\eta|<\delta_\varepsilon$, $n>n_\varepsilon$, we obtain

$$(2.54)\qquad \beta(\eta^-)-\varepsilon<\beta_n(\xi)<\beta(\eta^+)+\varepsilon.$$

Assume, at first, that $0<\eta<b$. We take δ'_ε with $0<\delta'_\varepsilon<\eta$, $\eta+\delta'_\varepsilon<b$, such that, when $|\xi-\eta|<\delta'_\varepsilon$, we have

$$\beta(\eta^-)-\varepsilon < \beta(\xi) < \beta(\eta^+)+\varepsilon. \tag{2.55}$$

We choose then $\delta_\varepsilon>0$ and n_ε such that $\delta_\varepsilon+(\theta/n_\varepsilon)<\delta'_\varepsilon$, $\eta+\delta_\varepsilon<\rho_{n_\varepsilon}$. Then, when $n>n_\varepsilon$, $|\xi-\eta|<\delta_\varepsilon$ $(\Rightarrow \xi<\rho_{n_\varepsilon})$,

$$\beta_n(\xi) = n\int_0^{\theta/n} \beta_n^+(\xi-s)\varphi(ns)ds = n\int_0^{\theta/n} \beta(\xi-s)\varphi(ns)ds. \tag{2.56}$$

Since

$$\eta-\delta'_\varepsilon < \eta-\delta_\varepsilon-\frac{\theta}{n} < \xi-s < \eta+\delta_\varepsilon < \eta+\delta'_\varepsilon, \tag{2.57}$$

from (2.55) and (2.56) follows (2.54).

In the same way it is possible to prove (2.54) when $\eta<0$.

Finally, let $\eta=0$. Then $\beta_n(0)=0$ and therefore, when $\xi>0$,

$$0 \le \beta_n(\xi) = n\int_0^{\theta/n} \beta_n^+(\xi-s)\varphi(ns)ds \le \beta(\xi^-). \tag{2.58}$$

Hence, if $0\le\xi<\delta_\varepsilon$,

$$0 \le \beta_n(\xi) < \beta(0^+)+\varepsilon. \tag{2.59}$$

The same procedure can be used to prove that, when $-\delta_\varepsilon<\xi\le 0$,

$$\beta(0^-)-\varepsilon < \beta_n(\xi) \le 0. \tag{2.60}$$

Then (2.54) holds also when $\eta=0$.

Let us now prove the existence theorem.

Consider the sequence $\{\beta_n(\eta)\}$ defined above and let $x_n(t)$ be the solution, in Γ, of the equation

$$Ax(t)-x''(t)+f(t) = \beta_n(x'(t)) \tag{2.61}$$

satisfying the initial conditions

$$x_n(0) = x_0, \qquad x'_n(0) = y_0. \tag{2.62}$$

This solution exists and is unique.

We obtain, moreover, by (2.40) and (2.47), since, by construction, $\|\beta_n(y_0)\|_{L^2}\le\|\bar\beta(y_0)\|_{L^2}$:

$$\begin{gathered}\|x_n(t)\|_{H_0^1} \le K_1, \qquad \|x'_n(t)\|_{H_0^1} \le K_2, \qquad \|Ax_n(t)\|_{L^2} \le K_2,\\ \|x''_n(t)\|_{L^2} \le 3\|\beta_n(y_0)\|_{L^2}+K_3 \le 3\|\bar\beta(y_0)\|_{L^2}+K_3,\\ \|\beta_n(x'_n(t))\|_{L^2} \le 3\|\beta_n(y_0)\|_{L^2}+K_4 \le 3\|\bar\beta(y_0)\|_{L^2}+K_4,\end{gathered} \tag{2.63}$$

where the quantities K_j do not depend on n.

By (2.63) we can assume, as before, that

$$\lim_{n\to\infty} x_n(t) \underset{E}{=} x(t) \qquad \text{uniformly on } 0 \vdash\!\dashv T, \tag{2.64}$$

$$\begin{aligned} \lim_{n\to\infty}{}^{*}\, x_n'(t) &\underset{L^2(0\vdash\dashv T;H_0^1)}{=} x'(t), \\ \lim_{n\to\infty}{}^{*}\, x_n''(t) &\underset{L^2(0\vdash\dashv T;L^2)}{=} x''(t), \\ \lim_{n\to\infty}{}^{*}\, Ax_n(t) &\underset{L^2(0\vdash\dashv T;L^2)}{=} Ax(t). \end{aligned} \tag{2.65}$$

We can also assume that

$$\lim_{n\to\infty} \frac{\partial x_n(t,\zeta)}{\partial t} = \frac{\partial x(t,\zeta)}{\partial t} \tag{2.66}$$

a.e. on Q and, by the last of (2.63),

$$\lim_{n\to\infty}{}^{*}\, \beta_n\left(\frac{\partial x_n(t,\zeta)}{\partial t}\right) \underset{L^2(Q)}{=} \chi(t,\zeta). \tag{2.67}$$

Therefore, $x(t) \in C^0(0\vdash\!\dashv T;E)$ and

$$\begin{aligned} &\|x'(t)\|_{H_0^1} \le K_2, \qquad \|Ax(t)\|_{L^2} \le K_2, \\ &\|x''(t)\|_{L^2} \le 3\|\bar{\beta}(y_0)\|_{L^2} + K_3. \end{aligned} \tag{2.68}$$

Hence, $x(t) \in \Gamma$ and $x'(t) \in L^\infty(0\vdash\!\dashv T;E)$, $Ax(t) \in L^\infty(0\vdash\!\dashv T;\, L^2)$.

By (2.67) and a theorem of Mazur there exists, $\forall n$, a sequence $\{\rho_{nk};\ k \ge n\}$, with $\rho_{nk} \ge 0$, $\sum_{(n)k}^{(\infty)} \rho_{nk} = 1$, such that

$$\lim_{n\to\infty} \sum_{n}^{\infty}{}_{k}\, \rho_{nk}\beta_k\left(\frac{\partial x_k(t,\zeta)}{\partial t}\right) \underset{L^2(Q)}{=} \chi(t,\zeta). \tag{2.69}$$

Hence, it is possible to select a subsequence $\{n_j\} \subseteq \{n\}$ such that, a.e. on Q,

$$\lim_{j\to\infty} \sum_{n_j}^{\infty}{}_{k}\, \rho_{n_jk}\beta_k\left(\frac{\partial x_k(t,\zeta)}{\partial t}\right) = \chi(t,\zeta). \tag{2.70}$$

Let $(\bar{t},\bar{\zeta})$ be a point in which (2.66) and (2.70) hold. Setting $\bar{\eta} = x_t(\bar{t},\bar{\zeta})$ and assuming $a < \bar{\eta} < b$, we shall prove that

$$\beta(\bar{\eta}^-) \le \chi(\bar{t},\bar{\zeta}) \le \beta(\bar{\eta}^+). \tag{2.71}$$

Chosen $\varepsilon > 0$ arbitrarily, let us determine δ_ε and n_ε' in such a way that, when $|\xi - \bar{\eta}| < \delta_\varepsilon$, $n > n_\varepsilon'$,

$$\beta(\bar{\eta}^-) - \varepsilon < \beta_n(\xi) < \beta(\bar{\eta}^+) + \varepsilon.$$

We now choose $n_\varepsilon > n_\varepsilon'$ such that, when $n > n_\varepsilon$,

$$\left| \frac{\partial x_n(\bar{t},\bar{\zeta})}{\partial t} - \frac{\partial x(\bar{t},\bar{\zeta})}{\partial t} \right| < \delta_\varepsilon.$$

Then

$$\beta(\bar{\eta}^-)-\varepsilon < \beta_n\left(\frac{\partial x_n(\bar{t},\bar{\zeta})}{\partial t}\right) < \beta(\bar{\eta}^+)+\varepsilon,$$

that is, always for $n > n_\varepsilon$,

$$\beta(\bar{\eta}^-)-\varepsilon < \sum_n^\infty {}_k\, \rho_{nk}\beta_k\left(\frac{\partial x_k(\bar{t},\bar{\zeta})}{\partial t}\right) < \beta(\bar{\eta}^+)+\varepsilon.$$

Hence, by (2.70),

$$\beta(\bar{\eta}^-)-\varepsilon \leq \chi(\bar{t},\bar{\zeta}) \leq \beta(\bar{\eta}^+)+\varepsilon,$$

which, since ε is arbitrary, proves (2.71).

Consider, finally, the case when $b < +\infty$, which implies

$$\lim_{\eta\to b^-} \beta(\eta) = +\infty.$$

Let Q_b be the set where $x_t(t,\zeta) \geq b$: it cannot be $m(Q_b)>0$. In fact, if it were so, there would exist, by (2.66), a set $Q_b' \subseteq Q_b$, with $m(Q_b')>0$, such that (2.66) holds uniformly on Q_b'. Since $x_t(t,\zeta)\geq b$, we can take, $\forall n$, $p_n \geq n$ such that

$$\frac{\partial x_{p_n}(t,\zeta)}{\partial t} \geq \rho_n, \qquad \forall (t,\zeta) \in Q_b'.$$

Hence,

$$\int_{Q_b'} \beta_{p_n}\left(\frac{\partial x_{p_n}(t,\zeta)}{\partial t}\right) dt d\zeta \geq \beta_{p_n}(\rho_n) m(Q_b') \to +\infty,$$

against the last of (2.63).

From (2.61), (2.64), (2.65), (2.71), it follows that $x(t)$ satisfies the equation

$$Ax(t)-x''(t)+f(t) = \beta(x'(t))$$

in the sense stated in § 1. Since $x(t)\in\Gamma$ and the initial conditions (1.9) are, by (2.62) and (2.64), obviously verified, the existence theorem is proved.

3. UNIQUENESS THEOREM OF THE BOUNDED SOLUTIONS

(*a*) Let $x(t)$ be a solution of (1.8) which is *bounded* on J, that is, such that (1.12) holds. We shall prove first that *such a solution is E-u.c. and has an E-r.c. range* $\mathscr{R}_{x(t)}$.

The E-uniform continuity of $x(t)$ follows, observing that we obtain, by (1.12), for $|\tau| \leq 1$,

$$\text{(3.1)} \qquad \|x(t+\tau)-x(t)\|_E \leq \left|\int_t^{t+\tau} \|x'(\eta)\|_E d\eta\right| \leq |\tau|^{1/2} M_x.$$

Assume now that the range $\mathscr{R}_{x(t)}$ is not r.c. There exist then a number $\rho > 0$ and a sequence $\{t_n\}$ such that

$$\|x(t_j) - x(t_k)\|_E \geq \rho \qquad (j \neq k). \tag{3.2}$$

Taking δ, with $0 < \delta \leq 1$, such that $\delta^{1/2} M_x \leq \rho/4$, consider the interval $t_n \vdash\!\dashv t_n + \delta$. Since, by (1.12),

$$\left\{\int_{t_n}^{t_n+\delta} (\|Ax(\eta)\|_{L^2}^2 + \|x'(\eta)\|_E^2) d\eta\right\}^{1/2} \leq M_x,$$

there exists $\eta_n \in t_n \vdash\!\dashv t_n + \delta$ such that

$$\|Ax(\eta_n)\|_{L^2} \leq M_x \delta^{-1/2}, \qquad \|x'(\eta_n)\|_{H_0^1} \leq M_x \delta^{-1/2}.$$

The sequences $\{x(\eta_n)\}$ and $\{x'(\eta_n)\}$ are, therefore, respectively H_0^1-r.c. and L^2-r.c. We can then select a subsequence (again denoted by $\{x(\eta_n)\}$) such that

$$\|x(\eta_j) - x(\eta_k)\|_E \leq \frac{\rho}{4}. \tag{3.3}$$

Since $\delta^{1/2} M_x \leq \rho/4$, it follows, by (3.1) and (3.3), that

$$\|x(t_j) - x(t_k)\|_E \leq \|x(t_j) - x(\eta_j)\|_E + \|x(\eta_j) - x(\eta_k)\|_E + \|x(\eta_k) - x(t_k)\|_E \leq \tfrac{3}{4}\rho,$$

which is absurd. Hence $\mathscr{R}_{x(t)}$ is r.c.

(*b*) We now prove theorem II.

Let $y(t)$ be another bounded solution and consider the sequences $\{x(t-n)\}$, $\{y(t-n)\}$ with $n = 0, 1, \cdots, t \in \Delta_p = -p \vdash\!\dashv p$ $(p = 1, 2, \cdots)$.

As $x(t)$ and $y(t)$ are E-u.c. and have E-r.c. ranges, it is possible, by the (vectorial) theorem of Ascoli-Arzelà, to select a subsequence $\{n_j\} \subseteq \{n\}$ such that

$$\lim_{j\to\infty} x(t - n_j) \underset{E}{=} \tilde{x}(t), \qquad \lim_{j\to\infty} y(t - n_j) \underset{E}{=} \tilde{y}(t) \tag{3.4}$$

uniformly on Δ_p, $\forall p$; moreover, by (1.12),

$$\begin{aligned} &\lim_{j\to\infty}{}^{*}\, x'(t-n_j) \underset{L^2(\Delta_p;E)}{=} \tilde{x}'(t), \qquad \lim_{j\to\infty}{}^{*}\, y'(t-n_j) \underset{L^2(\Delta_p;E)}{=} \tilde{y}'(t),\\ &\lim_{j\to\infty}{}^{*}\, Ax(t-n_j) \underset{L^2(\Delta_p;L^2)}{=} A\tilde{x}(t), \qquad \lim_{j\to\infty}{}^{*}\, Ay(t-n_j) \underset{L^2(\Delta_p;L^2)}{=} A\tilde{y}(t). \end{aligned} \tag{3.5}$$

By (1.13) we may assume that $(\forall p, Q_p = \Delta_p \times \Omega)$

$$\lim_{j\to\infty}{}^{*}\, f(t - n_j, \zeta) \underset{L^2(Q_p)}{=} g(t,\zeta) \tag{3.6}$$

and, by (3.4),

$$\lim_{j\to\infty} x_t(t - n_j, \zeta) \underset{L^2(Q_p)}{=} \tilde{x}_t(t,\zeta), \qquad \lim_{j\to\infty} y_t(t - n_j, \zeta) \underset{L^2(Q_p)}{=} \tilde{y}_t(t,\zeta). \tag{3.7}$$

Hence, we may assume that

$$\text{(3.8)} \qquad \lim_{j\to\infty} x_t(t-n_j,\,\zeta) = \tilde{x}_t(t,\zeta), \qquad \lim_{j\to\infty} y_t(t-n_j,\,\zeta) = \tilde{y}_t(t,\zeta)$$

a.e. on $J\times\Omega = Q_\infty$.

Observe that (since $\tilde{x}_{tt}(t,\zeta), \tilde{y}_{tt}(t,\zeta) \in L^2(Q_p), \forall p$), $\tilde{x}_t(t,\zeta)$ and $\tilde{y}_t(t,\zeta)$ are continuous functions of $t\in J$ for all $\zeta\in\Omega$, with the exception of those belonging to a set of measure zero. Furthermore, by (3.4), $\overline{\mathscr{R}}_{\tilde{x}(t)} \le \overline{\mathscr{R}}_{x(t)}$ on E and, by (3.5),

$$\text{(3.9)} \qquad \left\{\int_t^{t+1} (\|A\tilde{x}(\eta)\|_{L^2}^2 + \|\tilde{x}'(\eta)\|_E^2)d\eta\right\}^{1/2} \le M_x.$$

Analogous relations hold for $\tilde{y}(t)$.

Moreover, a.e. on Q_∞,

$$\text{(3.10)} \qquad a < \tilde{x}_t(t,\zeta) < b, \qquad a < \tilde{y}_t(t,\zeta) < b$$

and we may assume that, $\forall Q_p$,

$$\text{(3.11)} \qquad \begin{aligned} &\lim_{j\to\infty}{}^{*}\, \beta(x_t(t-n_j,\,\zeta)) \underset{L^2(Q_p)}{=} \chi_1(t,\zeta),\\ &\lim_{j\to\infty}{}^{*}\, \beta(y_t(t-n_j,\,\zeta)) \underset{L^2(Q_p)}{=} \chi_2(t,\zeta), \end{aligned}$$

where

$$\text{(3.12)} \qquad \begin{aligned} &\chi_1(t,\zeta) \in \beta((\tilde{x}_t(t,\zeta))^-) \vdash\!\dashv \beta((\tilde{x}_t(t,\zeta))^+),\\ &\chi_2(t,\zeta) \in \beta((\tilde{y}_t(t,\zeta))^-) \vdash\!\dashv \beta((\tilde{y}_t(t,\zeta))^+). \end{aligned}$$

The proof of (3.10), (3.11), and (3.12) is analogous to that given at § 2 (*c*) of this chapter and we shall therefore omit it here.

Hence, $\tilde{x}(t)$ and $\tilde{y}(t)$ are bounded solutions, on J, of the equation

$$Au(t) - u''(t) + g(t) = \beta(u'(t)).$$

Let us prove now that $x(t) = y(t)$.

If we set $w(t) = y(t) - x(t)$, the function $w(t)$ obviously satisfies the equation

$$w''(t) - Aw(t) = -\{\beta(x'(t) + w'(t)) - \beta(x'(t))\}.$$

Hence, if $t_1 < t_2$,

$$\|w(t_2)\|_E^2 = \|w(t_1)\|_E^2 - 2\int_{t_1}^{t_2} (\beta(x'(t) + w'(t)) - \beta(x'(t)), w'(t))_{L^2} dt.$$

The function $\|w(t)\|_E$ is therefore decreasing and, as it is nonnegative and bounded, there exist the limits

$$\lim_{t\to+\infty} \|w(t)\|_E = N_1 < +\infty, \qquad \lim_{t\to-\infty} \|w(t)\|_E = N_2 < +\infty,$$

where

$$\text{(3.13)} \qquad 0 \le N_2^2 - N_1^2 = 2\int_J (\beta(x'(t) + w'(t)) - \beta(x'(t)), w'(t))_{L^2} dt < +\infty.$$

Consider the sequence $\{w(t-n)\}$, $n=0, 1, \cdots$. We have, by (3.4), (3.5), (3.7), (3.8),

$$\lim_{j\to\infty} w(t-n_j) \underset{E}{=} \tilde{w}(t) = \tilde{y}(t)-\tilde{x}(t),$$

$$\lim_{j\to\infty}{}^{*}\, w'(t-n_j) \underset{L^2(\Delta_p;E)}{=} \tilde{w}'(t), \qquad \lim_{j\to\infty}{}^{*}\, Aw(t-n_j) \underset{L^2(\Delta_p;L^2)}{=} A\tilde{w}(t);$$

moreover, a.e. on Q_∞,

$$\lim_{j\to\infty} w_t(t-n_j,\zeta) = \tilde{w}_t(t,\zeta). \tag{3.14}$$

We shall prove, first, that $\tilde{w}_t(t,\zeta)=0$ almost everywhere on Q_∞.

Assume, in fact, that $\tilde{w}_t(t,\zeta)>0$ on a set $\tilde{Q}$ with $m(\tilde{Q})>0$.

We can, obviously, suppose that $\tilde{Q}$ is closed and bounded, that the functions $x_t(t-n_j,\zeta)$, $y_t(t-n_j,\zeta)$ are continuous on $\tilde{Q}$, and that the convergence corresponding to (3.8) is uniform.

Moreover, we can assume that $\tilde{Q}\subseteq(k \vdash\!\!\dashv k+1)\times\Omega$, where k is a suitable integer. Therefore, on $\tilde{Q}$,

$$\tilde{w}_t(t,\zeta) = \tilde{y}_t(t,\zeta)-\tilde{x}_t(t,\zeta) \geq 2\rho > 0$$

and, consequently, when $j\geq\bar{j}$,

$$y_t(t-n_j,\zeta)-x_t(t-n_j,\zeta) \geq \rho > 0, \tag{3.15}$$

$$|y_t(t-n_j,\zeta)| \leq M < +\infty, \qquad |x_t(t-n_j,\zeta)| \leq M < +\infty.$$

Let us consider the function $\beta(\eta)$. We obtain ($\forall\xi,\eta\in a \frown b$)

$$\beta(\eta)-\beta(\xi) \geq \beta(\eta^-)-\beta(\xi^+).$$

Hence, if $\eta\geq\xi+\rho$ (and assuming $\rho<b-a$)

$$\beta(\eta)-\beta(\xi) \geq \beta((\xi+\rho)^-)-\beta(\xi^+) > 0.$$

Let $|\xi|\leq M$, $|\eta|\leq M$, $\eta-\xi\geq\rho$. Then

$$\beta((\xi+\rho)^-)-\beta(\xi^+) \geq \sigma > 0. \tag{3.16}$$

In fact, if (3.16) did not hold, there would exist a sequence $\{\xi_n\}$ such that $\lim_{n\to\infty}\xi_n=\bar{\xi}\in a \frown b$, $\xi_n\neq\bar{\xi}$, and

$$\lim_{n\to\infty}\{\beta((\xi_n+\rho)^-)-\beta(\xi_n^+)\} = 0. \tag{3.17}$$

However, if $\xi_n\downarrow\bar{\xi}$,

$$\lim_{n\to\infty}\{\beta((\xi_n+\rho)^-)-\beta(\xi_n^+)\} = \beta((\bar{\xi}+\rho)^+)-\beta(\bar{\xi}^+) > 0,$$

whereas, if $\xi_n\uparrow\bar{\xi}$,

$$\lim_{n\to\infty}\{\beta((\xi_n+\rho)^-)-\beta(\xi_n^+)\} = \beta((\bar{\xi}+\rho)^-)-\beta(\bar{\xi}^-) > 0.$$

Relation (3.17) is therefore absurd and (3.16) holds.

Hence, by (3.13), (3.15), and (3.16),

$$N_2^2 - N_1^2 = 2\int_{-\infty}^{+\infty} dt \int_{\Omega} (\beta(y_t(t,\zeta)) - \beta(x_t(t,\zeta))(y_t(t,\zeta) - x_t(t,\zeta))d\zeta$$

$$\geq 2\sum_j^{\infty} \int_{\tilde{Q}} (\beta(y_t(t-n_j,\zeta)) - \beta(x_t(t-n_j,\zeta)))$$
$$\times (y_t(t-n_j,\zeta) - x_t(t-n_j,\zeta))dtd\zeta$$

$$\geq 2\sum_j^{\infty} \rho\sigma m(\tilde{Q}) = +\infty,$$

which contradicts (3.13). It cannot therefore be $\tilde{w}_t(t,\zeta) > 0$ on a set of positive measure.

In the same way it can be shown that $\tilde{w}_t(t,\zeta)$ cannot be <0 on a set of positive measure. Therefore, since

$$\tilde{w}_t(t,\zeta) = 0 \qquad \text{a.e. on } Q_\infty, \tag{3.18}$$

$\tilde{w}(t,\zeta)$ does not depend on t: $\tilde{w}(t,\zeta) = \tilde{w}(\zeta)$.

Moreover, from (3.18) it follows that $\tilde{w}_{tt}(t,\zeta) = 0$.

The function $\tilde{w}(t,\zeta)$, on the other hand, satisfies, by what has been proved above, the equation

$$A(\zeta)\tilde{w}(t,\zeta) - \tilde{w}_{tt}(t,\zeta) = \chi_1(t,\zeta) - \chi_2(t,\zeta)$$

and, consequently,

$$A(\zeta)\tilde{w}(\zeta) = \chi_1(t,\zeta) - \chi_2(t,\zeta) = h(\zeta). \tag{3.19}$$

We shall now prove that $h(\zeta) = 0$, a.e. on Ω.

As previously observed, $\tilde{x}_t(t,\zeta)$, $\tilde{y}_t(t,\zeta)$ are continuous functions of t, $\forall \zeta \in \Omega_0 \subset \Omega$ with $m(\Omega_0) = m(\Omega)$. Let us take $\bar{\zeta} \in \Omega_0$; if there exists $\bar{t} \in J$ such that $\beta(\eta)$ is continuous at $\bar{\eta} = \tilde{x}_t(\bar{t},\bar{\zeta}) = \tilde{y}_t(\bar{t},\bar{\zeta})$, then, by (3.12),

$$h(\bar{\zeta}) = \chi_1(\bar{t},\bar{\zeta}) - \chi_2(\bar{t},\bar{\zeta}) = 0.$$

Otherwise, $\eta = \tilde{x}_t(t,\bar{\zeta})$ will be a point at which β is not continuous, $\forall t \in J$.

Suppose now $h(\bar{\zeta}) \neq 0$. There cannot exist two values t_1 and t_2 such that

$$\eta_1 = \tilde{x}_t(t_1,\bar{\zeta}) < \tilde{x}_t(t_2,\bar{\zeta}) = \eta_2,$$

$\beta(\eta)$ being discontinuous both at η_1 and η_2. In fact, as $\tilde{x}_t(t,\bar{\zeta})$ is a continuous function of t and β is strictly increasing, it would be $\beta(\eta_1^+) < \beta(\eta_2^-)$ and, if $\eta' \in \overline{\eta_1 \eta_2}$ is a point at which β is continuous, there exists $t' \in \overline{t_1 t_2}$ such that $\tilde{x}_t(t',\bar{\zeta}) = \eta'$. Hence,

$$h(\bar{\zeta}) = \chi_1(t',\bar{\zeta}) - \chi_2(t',\bar{\zeta}) = 0,$$

which is in contrast with our assumption.

Therefore, $\forall t \in J$, $\tilde{x}_t(t,\zeta) = \tilde{x}_t(t_1,\zeta) = \eta_1$, $\beta(\eta)$ being discontinuous at η_1. Hence, on all J,

$$\tilde{x}(t,\zeta) = \tilde{x}(0,\zeta) + t\tilde{x}_t(0,\zeta), \tag{3.20}$$

which implies necessarily (as $\tilde{x}(t)$ is L^2-bounded) $\tilde{x}_t(0,\zeta) = 0$, $\eta_1 = 0$.

Since $\beta(\eta)$ is continuous at $\eta = 0$, it follows again that $h(\zeta) = 0$. Then, by (3.19), a.e. on Ω,

$$A(\zeta)\tilde{w}(\zeta) = 0,$$

which implies

$$\tilde{w}(\zeta) = 0.$$

Since

$$\|\tilde{w}\|_E = \lim_{j\to\infty} \|w(t-n_j)\|_E = N_2,$$

it follows that $N_2 = 0 \Rightarrow N_1 = 0$. Hence, $\|w(t)\|_E = 0$, that is, $x(t) = y(t)$, which proves the theorem.

Observation III. Theorem II can be generalized. Assuming, in fact, that

$$\max \lim_{t\to-\infty} \|f(t)\|_{L^2(L^2)} < +\infty,$$

one proves that there exists at most one solution $x(t)$, $t \in J$, such that

$$\max \lim_{t\to-\infty} \{\|Ax(t)\|^2_{L^2(L^2)} + \|x'(t)\|^2_{L^2(E)}\}^{1/2} < +\infty.$$

4. ALMOST-PERIODICITY THEOREM

Proof. Let $\{s_n\}$ be any real sequence. By (1.12) it is possible, as in theorem II, to select from $\{s_n\}$ a subsequence (which will again be denoted by $\{s_n\}$) such that, $\forall t \in J$,

$$\begin{aligned} &\lim_{n\to\infty} x(t+s_n) \underset{E}{=} z(t), \\ &\lim_{n\to\infty}{}^{*}\, x'(t+s_n) \underset{L^2(E)}{=} z'(t), \\ &\lim_{n\to\infty}{}^{*}\, Ax(t+s_n) \underset{L^2(L^2)}{=} Az(t). \end{aligned} \tag{4.1}$$

We may, moreover, obviously assume that

$$\lim_{n\to\infty}{}^{*}\, f(t+s_n) \underset{L^2(L^2)}{=} g(t) \tag{4.2}$$

uniformly on J.

Regarding (4.1) and (4.2), we recall that, for instance, the second equation of (4.1) means:

$$\lim_{n\to\infty}\int_0^1\{(x'(t+s_n+\eta),g(\eta))_{H_0^1}+(x''(t+s_n+\eta),g'(\eta))_{L^2}\}d\eta$$
$$=\int_0^1\{(z'(t+\eta),g(\eta))_{H_0^1}+(z''(t+\eta),g'(\eta))_{L^2}\}d\eta,$$

$\forall g\in L^2(H_0^1)$, with $g'\in L^2(L^2)$.

The function $z(t)$ is a bounded solution, on J, of the equation

$$Az(t)-z''(t)+g(t)=\beta(z'(t)).$$

In order to prove the theorem, it will be sufficient, by Bochner's criterion, to show that (4.1) hold uniformly on J. Assume, for instance, that this does not occur for the first of (4.1). There exists then a number $\sigma>0$ and three sequences $\{t_n\}$, $\{s_{n1}\}\subseteq\{s_n\}$, $\{s_{n2}\}\subseteq\{s_n\}$ such that

$$\|x(t_n+s_{n1})-x(t_n+s_{n2})\|_E\geq\sigma. \tag{4.3}$$

We can, on the other hand, select from $\{t_n+s_{n1}\}$, $\{t_n+s_{n2}\}$ two subsequences (which will be denoted by the same notations) such that, uniformly on J,

$$\lim_{n\to\infty}{}^{*}\, f(t+t_n+s_{nk})\underset{L^2(L^2)}{=}g_k(t)\qquad(k=1,2) \tag{4.4}$$

and, moreover, $\forall t\in J$,

$$\begin{aligned}
&\lim_{n\to\infty}x(t+t_n+s_{n1})\underset{E}{=}z_1(t), &&\lim_{n\to\infty}x(t+t_n+s_{n2})\underset{E}{=}z_2(t),\\
&\lim_{n\to\infty}{}^{*}\,x'(t+t_n+s_{n1})\underset{L^2(E)}{=}z_1'(t), &&\lim_{n\to\infty}{}^{*}\,x'(t+t_n+s_{n2})\underset{L^2(E)}{=}z_2'(t),\\
&\lim_{n\to\infty}{}^{*}\,Ax(t+t_n+s_{n1})\underset{L^2(L^2)}{=}Az_1(t), &&\lim_{n\to\infty}{}^{*}\,Ax(t+t_n+s_{n2})\underset{L^2(L^2)}{=}Az_2(t).
\end{aligned} \tag{4.5}$$

Let us observe now that, on all J,

$$g_1(t)=g_2(t). \tag{4.6}$$

In fact, by (4.2) and (4.4),

$$\|f(t+t_n+s_{n1})-f(t+t_n+s_{n2})\|_{L^2(L^2)}$$
$$\leq\|f(t+t_n+s_{n1})-f(t+t_n+s_n)\|_{L^2(L^2)}$$
$$+\|f(t+t_n+s_n)-f(t+t_n+s_{n2})\|_{L^2(L^2)}\to 0.$$

The limit functions $z_1(t)$ and $z_2(t)$ are then bounded solutions of the equation

$$Az(t)-z''(t)+g_1(t)=\beta(z'(t)).$$

By the uniqueness theorem of the bounded solution, it follows that $z_1(t)=z_2(t)$: in particular, $z_1(0)=z_2(0)$, which contradicts (4.3). Hence the convergence corresponding to the first of (4.1) is uniform on J, and $x(t)$ is E-a.p.

In the same way it can be shown that the second and third equations of (4.1) hold uniformly on J; consequently, $x'(t)$ and $Ax(t)$ are E-S^2 w.a.p. and L^2-S^2 w.a.p., respectively.

5. ANALYSIS OF A SPECIAL CASE

(a) *Proof of theorem IV.* Let us first prove the existence of the solution. Let $\{g_j\}$ be a basis of $H_0^1 \cap L^p$ and set, as in § 2,

$$x_n(t) = \sum_1^n {}_j\, \alpha_{n_j}(t) g_j.$$

Consider the system of "approximating equations"

$$(5.1) \quad (x_n''(t),g_j)_{L^2} - (Ax_n(t),g_j)_{L^2} + \langle \beta(x_n'(t),g_j\rangle = \langle f(t),g_j\rangle \quad (j = 1,\cdots,n)$$

with the initial conditions

$$(5.2) \qquad x_n(0) = x_{0,n}, \qquad x_n'(0) = y_{0,n},$$

where $x_{0,n}$, $y_{0,n}$ are chosen in such a way that $x_{0,n} \underset{H_0^1}{\rightarrow} x_0$, $y_{0,n} \underset{L^2}{\rightarrow} y_0$, being moreover $\|x_{0,n}\|_{H_0^1} \leq \|x_0\|_{H_0^1}$, $\|y_{0,n}\|_{L^2} \leq \|y_0\|_{L^2}$.

As is well known, the solution of (5.1) and (5.2) exists in some interval $0 \leq t \leq \delta_n$, $\delta_n \leq T$.

Multiplying (5.1) by $\alpha_{nj}'(t)$, adding and integrating between 0 and $t \in 0 \vdash\!\dashv \delta_n$, we obtain

$$(5.3) \quad \|x_n(t)\|_E^2 + 2\int_0^t \langle \beta(x_n'(\eta)),x_n'(\eta)\rangle d\eta = \|x_n(0)\|_E^2 + 2\int_0^t \langle f(\eta),x_n'(\eta)\rangle d\eta.$$

By (1.14) we have, on the other hand, $\forall v \in L^p$,

$$(5.4) \qquad \int_\Omega v(\zeta)\beta(v(\zeta))d\zeta \geq c_1 \int_\Omega |v(\zeta)|^p d\zeta - K_1,$$

where K_1 depends only on Ω, p, $\bar{\eta}$.

Hence, by (5.3) and (5.4),

$$(5.5) \quad \|x_n(t)\|_E^2 + 2c_1 \int_0^t \|x_n'(\eta)\|_{L^p}^p d\eta$$
$$\leq \|x_n(0)\|_E^2 + 2K_1 t + 2\left\{\int_0^t \|f(\eta)\|_{L^r}^r d\eta\right\}^{1/r} \left\{\int_0^t \|x_n'(\eta)\|_{L^p}^p d\eta\right\}^{1/p}.$$

We have moreover

$$(5.6)\qquad \|x_n(t)\|_E^2+c_1\int_0^t\|x_n'(\eta)\|_{L^p}^p d\eta \le \|x_n(0)\|_E^2+K_2\int_0^t\|f(\eta)\|_{L^r}^r d\eta+2K_1 t,$$

K_2 being independent of n.

In fact, if

$$\int_0^t\|x_n'(\eta)\|_{L^p}^p d\eta \le \left(\frac{2}{c_1}\right)^r\int_0^t\|f(\eta)\|_{L^r}^r d\eta,$$

then relation (5.6) holds, with $K_2=2(2/c_1)^{r/p}$. If, instead,

$$\int_0^t\|x_n'(\eta)\|_{L^p}^p d\eta > \left(\frac{2}{c_1}\right)^r\int_0^t\|f(\eta)\|_{L^r}^r d\eta,$$

we have, by (5.5),

$$\|x_n(t)\|_E^2+2c_1\int_0^t\|x_n'(\eta)\|_{L^p}^p d\eta \le \|x_n(0)\|_E^2+c_1\int_0^t\|x_n'(\eta)\|_{L^p}^p d\eta+2K_1 t.$$

Hence (5.6) holds in any case.

Since, by (5.2), $\|x_n(0)\|_E^2\le\|x_0\|_{H_0^1}^2+\|y_0\|_{L^2}^2$, it follows from (5.6) that $\delta_n=T$ and that

$$(5.7)\qquad \sup_{0\le t\le T}\|x_n(t)\|_E\le K_3<+\infty,\qquad \int_0^T\|x_n'(t)\|_{L^p}^p dt\le K_4<+\infty,$$

where K_3 and K_4 do not depend on n.

Observe that, from (1.14) it follows, when $|\eta|\ge\bar\eta$, that $|\beta(\eta)|^r\le c_2|\eta|^{(p-1)r}=c_2|\eta|^p$; therefore, by the second of (5.7),

$$(5.8)\qquad \int_0^T\|\beta(x_n'(t))\|_{L^r}^r dt\le c_2\int_0^T\|x_n'(t)\|_{L^p}^p dt+MT,$$

where $M=\max\limits_{|\eta|\le\bar\eta}|\beta(\eta)|^2 m(\Omega)$.

We may then select (by (5.7) and (5.8)) from $\{x_n(t)\}$ a subsequence (again denoted by $\{x_n(t)\}$) such that

$$(5.9)\qquad \begin{aligned}\lim_{n\to\infty}{}^*\, x_n(t) &\underset{L^2(0\mapsto T;E)}{=} x(t),\\ \lim_{n\to\infty}{}^*\, x_n'(t) &\underset{L^p(0\mapsto T;L^p)}{=} x'(t),\\ \lim_{n\to\infty}{}^*\, \beta(x_n'(t)) &\underset{L^r(0\mapsto T;L^r)}{=} \chi(t).\end{aligned}$$

Moreover, we have

$$\sup_{0\le t\le T}\|x(t)\|_E\le K_3.$$

The limit function therefore satisfies condition (i_1').

Let us prove that $x(t)$ is a solution of the equation

$$x''(t) - Ax(t) + \chi(t) = f(t). \tag{5.10}$$

In fact, by (5.9),

$$\lim_{n\to\infty} \int_0^T \{(x_n''(t),h(t))_{L^2} + (x_n(t),h(t))_{H_0^1} + \langle\beta(x_n'(t)),h(t)\rangle - \langle f(t),h(t)\rangle\}dt$$

$$\begin{aligned}(5.11)\qquad &= \lim_{n\to\infty}\int_0^T \{-(x_n'(t),h'(t))_{L^2} + (x_n(t),h(t))_{H_0^1} \\ &\qquad + \langle\beta(x_n'(t)),h(t)\rangle - \langle f(t),h(t)\rangle\}dt \\ &= \int_0^T \{-(x'(t),h'(t))_{L^2} + (x(t),h(t))_{H_0^1} + \langle\chi(t),h(t)\rangle - \langle f(t),h(t)\rangle\}dt,\end{aligned}$$

$\forall h(t) \in L^1(0 \vdash\!\dashv T; E) \cap L^p(0 \vdash\!\dashv T; L^p)$ with compact support on $0 \overline{\quad} T$.

The "approximate solutions" $x_n(t)$ satisfy, on the other hand, by (5.1), the equation

$$\int_0^T \{(x_n''(t),l(t))_{L^2} + (x_n(t), l(t))_{H_0^1} + \langle\beta(x_n'(t)),l(t)\rangle - \langle f(t),l(t)\rangle\}dt = 0, \tag{5.12}$$

$\forall l(t) = \sum_{(1)j}^{(m)} \omega_j(t) g_j$ (m fixed $\leq n$) with $\omega_j(t) \in \mathscr{D}(0 \overline{\quad} T)$.

As the space of functions $l(t)$ so defined is dense in that of the test functions $h(t)$, from (5.11) and (5.12) it follows that $x(t)$ is a solution of (5.10) on $0 \vdash\!\dashv T$, $\forall T > 0$.

To complete the proof of the existence theorem, we have now to show that $\chi(t,\zeta) = \beta(\partial x(t,\zeta)/\partial t)$ a.e. on $Q = 0 \vdash\!\dashv T \times \Omega$ and that the initial conditions $x(0) = x_0$, $x'(0) = y_0$ are satisfied.

This can be done following the procedure given by Lions and Strauss.

We shall, however, give here a simpler proof, *under the additional assumption that $x_0 = y_0 = 0$.*

We shall in fact utilize essentially the present theorem in this more particular form. For the proof in the general case we refer therefore to Lions and Strauss ([1], part 2, theorem 2.1).

By (5.3) and the second of (5.7) we have, taking $x_n(0) = x_n'(0) = 0$,

$$\|x_n(t)\|_E^2 \leq 2\int_0^t \langle f(\eta),x_n'(\eta)\rangle d\eta \leq 2K_4^{1/p}\left\{\int_0^t \|f(\eta)\|_{L^r}^r d\eta\right\}^{1/r} = \psi(t), \tag{5.13}$$

with $\lim_{t\to 0} \psi(t) = 0$.

Consider now the difference $x_n(t+\delta) - x_n(t)$ ($\delta > 0$), which obviously satisfies the system, analogous to (5.1),

$$\begin{aligned}(x_n''(t+\delta) - x_n''(t),g_j)_{L^2} - (Ax_n(t+\delta) - Ax_n(t),g_j)_{L^2} + \langle\beta(x_n'(t+\delta)) - \beta(x_n'(t)),g_j\rangle \\ = \langle f(t+\delta) - f(t),g_j\rangle \qquad (j = 1,\cdots,n).\end{aligned}$$

Hence, bearing in mind that $[\beta(\xi_2)-\beta(\xi_1)](\xi_2-\xi_1)\geq 0$, we have

$$\begin{aligned}\|x_n(t+\delta)-x_n(t)\|_E^2 &\leq \|x_n(\delta)-x_n(0)\|_E^2 \\ &\quad +2\int_0^t \langle f(\eta+\delta)-f(\eta), x_n'(\eta+\delta)-x_n'(\eta)\rangle d\eta \\ &= \|x_n(\delta)\|_E^2+2\int_0^t \langle f(\eta+\delta)-f(\eta), x_n'(\eta+\delta)-x_n'(\eta)\rangle d\eta.\end{aligned} \tag{5.14}$$

By (5.7) and (5.14) and observing that $f(t)$ is $L^r(L^r)$-continuous, we obtain, for $0\leq t\leq T$,

$$\begin{aligned}\|x_n(t+\delta)-x_n(t)\|_E^2 &\leq \psi(\delta)+2\left\{\int_0^T \|f(\eta+\delta)-f(\eta)\|_{L^r}^r d\eta\right\}^{1/r} \\ &\quad \times\left\{\int_0^t \|x_n'(\eta+\delta)-x_n'(\eta)\|_{L^p}^p d\eta\right\}^{1/p} \\ &\leq \psi(\delta)+K_5\varphi(\delta),\end{aligned} \tag{5.15}$$

being $\varphi(\delta)=\sup_{0\vdash\dashv T}\|f(t+\delta)-f(t)\|_{L^r(L^r)}$, $\lim_{\delta\to 0}\varphi(\delta)=0$.

Hence the sequence $\{x_n(t)\}$ is E-equally continuous on $0\vdash\dashv T$.

As the embedding of H_0^1 in L^2 is compact, the values $x_n(t)$ have, $\forall$ fixed $t\in 0\vdash\dashv T$, an L^2-r.c. range. Consequently, by the (vectorial) theorem of Ascoli-Arzelà, we may assume that

$$\lim_{n\to\infty} x_n(t) \underset{L^2}{=} x(t) \tag{5.16}$$

uniformly on $0\vdash\dashv T$.

Observe now that, as the functions $x_n'(t)$ are L^2-equally continuous on $0\vdash\dashv T$, from (5.16) follows (by a known theorem of differentiation by series), uniformly on $0\vdash\dashv T$,

$$\lim_{n\to\infty} x_n'(t) \underset{L^2}{=} x'(t). \tag{5.17}$$

It is therefore possible to assume that

$$\lim_{n\to\infty}\frac{\partial x_n(t,\zeta)}{\partial t}=\frac{\partial x(t,\zeta)}{\partial t} \quad \text{a.e. on } Q. \tag{5.18}$$

Relation (5.18) implies that

$$\lim_{n\to\infty}\beta\left(\frac{\partial x_n(t,\zeta)}{\partial t}\right)=\beta\left(\frac{\partial x(t,\zeta)}{\partial t}\right) \quad \text{a.e. on } Q. \tag{5.19}$$

Chosen an arbitrary $\varepsilon>0$, there exists then a closed set $Q_\varepsilon\subset Q$, with $m(Q-Q_\varepsilon)<\varepsilon$, such that in Q_ε the convergence in (5.18) is uniform and all the functions $\partial x_n(t,\zeta)/\partial t$ are continuous.

$\forall v \in L^p(Q)$, with $v=0$ on $Q-Q_\varepsilon$, we obtain therefore, bearing in mind the last of (5.9),

$$\lim_{n\to\infty}\int_Q \beta\left(\frac{\partial x_n(t,\zeta)}{\partial t}\right)v(t,\zeta)dQ = \int_{Q_\varepsilon}\chi(t,\zeta)v(t,\zeta)dQ_\varepsilon = \int_{Q_\varepsilon}\beta\left(\frac{\partial x(t,\zeta)}{\partial t}\right)v(t,\zeta)dQ_\varepsilon. \tag{5.20}$$

Since v is arbitrary, it follows that

$$\chi(t,\zeta) = \beta\left(\frac{\partial x(t,\zeta)}{\partial t}\right)$$

a.e. on Q_ε and therefore also on Q.

Regarding the initial conditions, we have obviously, by (5.16) and (5.17),

$$x(0) \underset{L^2}{=} 0, \qquad x'(0) \underset{L^2}{=} 0.$$

From (5.15) we obtain moreover, letting $n\to\infty$,

$$\|x(t+\delta)-x(t)\|_E^2 \le \psi(\delta)+K_5\varphi(\delta),$$

that is, $x(t)$ *is E-continuous.*

Let us now prove the uniqueness of the solution.

Let $x(t)$, $y(t)$ be two solutions satisfying the same initial conditions. If we set $w(t)=x(t)-y(t)$, the function $w(t)$ satisfies the equation, analogous to (1.20),

$$\|w(t)\|_E^2 = \|w(0)\|_E^2 - 2\int_0^t \langle\beta(x'(\eta))-\beta(y'(\eta)),x'(\eta)-y'(\eta)\rangle d\eta,$$

from which follows, since $\|w(0)\|_E=0$ and

$$\int_0^t \langle\beta(x'(\eta))-\beta(y'(\eta),x'(\eta))-y'(\eta)\rangle d\eta \ge 0,$$

that $\|w(t)\|_E=0$.

Observation IV. From (1.20) we obtain, by (1.14), $\forall t_2>t_1$,

$$\|x(t_2)\|_E^2 + c_1\int_{t_1}^{t_2}\|x'(\eta)\|_{L^p}^p d\eta \le \|x(t_1)\|_E^2 + K_2\int_{t_1}^{t_2}\|f(\eta)\|_{L^r}^r d\eta + 2K_1(t_2-t_1). \tag{5.21}$$

(*b*) *Proofs of theorems V and VI.* Since the proofs are very similar, we shall give only the second one. We have, by assumption,

$$\sup_{J_0}\|f(t+\delta)-f(t)\|_{L^r(L^r)} = \sup_{J_0}\left\{\int_0^1 \|f(t+\delta+\eta)-f(t+\eta)\|_{L^r}^r d\eta\right\}^{1/r} = \varphi_0(\delta) \qquad (\delta>0) \tag{5.22}$$

where

$$\lim_{\delta\to 0}\varphi_0(\delta) = 0. \tag{5.23}$$

Observe now, first of all, that, if $x_1(t)$ and $x_2(t)$ are two solutions, corresponding respectively to the known terms $f_1(t)$ and $f_2(t)$, the function $w(t)=x_1(t)-x_2(t)$ is a solution of the equation

$$Aw(t)-w''(t)+f_1(t)-f_2(t) = \beta(x_1'(t))-\beta(x_2'(t)). \tag{5.24}$$

The function $w(t)$ moreover satisfies a relation analogous to (1.20); we have then, by (1.22), for $t_1 < t_2$,

$$\|w(t_2)\|_E^2+2c_3\int_{t_1}^{t_2}\|w'(t)\|_{L^p}^p dt \le \|w(t_1)\|_E^2+2\int_{t_1}^{t_2}\langle f_1(t)-f_2(t),w'(t)\rangle dt. \tag{5.25}$$

Multiplying (5.24) by $w(t)$ and integrating on $t_1 \vdash\!\dashv t_2$, we obtain

$$\begin{aligned}(w'(t_2),w(t_2))_{L^2}-(w'(t_1),w(t_1))_{L^2} \\ +\int_{t_1}^{t_2}\{-\|w'(t)\|_{L^2}^2+\|w(t)\|_{H_0^1}^2+\langle\beta(x_1'(t))-\beta(x_2'(t)),w(t)\rangle \\ -\langle f_1(t)-f_2(t),w(t)\rangle\}dt = 0.\end{aligned} \tag{5.26}$$

In what follows we shall denote by $\varphi_j(\delta)$ continuous functions, ≥ 0 on $0 \vdash\!\dashv 1$ such that $\varphi_j(0)=0$ and by c_j positive constants.

Since $x(t)$ is E-continuous, we have, setting $w_\delta(t) = x(t+\delta)-x(t)$,

$$\sup_{0\leqslant t\leqslant 1}\|w_\delta(t)\|_E = \varphi_1(\delta). \tag{5.27}$$

Applying relation (5.25) to $w_\delta(t)$, we obtain, $\forall \bar t\geq 0$,

$$\begin{aligned}\|w_\delta(\bar t+1)\|_E^2-\|w_\delta(\bar t)\|_E^2 \\ \le 2(-c_3\|w_\delta'(\bar t)\|_{L^p(L^p)}^p+\|f(\bar t+\delta)-f(\bar t)\|_{L^r(L^r)}\|w_\delta'(\bar t)\|_{L^p(L^p)}) \\ = 2\|w_\delta'(\bar t)\|_{L^p(L^p)}(-c_3\|w_\delta'(\bar t)\|_{L^p(L^p)}^{p-1}+\|f(\bar t+\delta)-f(\bar t)\|_{L^r(L^r)}).\end{aligned} \tag{5.28}$$

Assume, at first, that

$$\|w_\delta'(\bar t)\|_{L^p(L^p)}^{p-1} > \frac{1}{c_3}\varphi_0(\delta). \tag{5.29}$$

It follows then from (5.28) that

$$\|w_\delta(\bar t+1)\|_E \le \|w_\delta(\bar t)\|_E. \tag{5.30}$$

If, on the contrary, (5.29) does not hold, but instead

$$\|w_\delta'(\bar t)\|_{L^p(L^p)}^{p-1} \le \frac{1}{c_3}\varphi_0(\delta), \tag{5.31}$$

then, denoting by c_5 the embedding constant of $L^p(L^p)$ in $L^2(L^2)$, we have

$$\|w_\delta'(\bar t)\|_{L^2(L^2)} \le c_5\|w_\delta'(\bar t)\|_{L^p(L^p)} \le c_5\left(\frac{\varphi_0(\delta)}{c_3}\right)^{1/(p-1)} = \varphi_2(\delta). \tag{5.32}$$

Hence, there exist two points $t'_\delta \in \bar{t} \mapsto \bar{t}+\frac{1}{4}$, $t''_\delta \in \bar{t}+\frac{3}{4} \mapsto \bar{t}+1$ such that

$$\|w'_\delta(t'_\delta)\|_{L^2} \leq 4\varphi_2(\delta), \qquad \|w'_\delta(t''_\delta)\|_{L^2} \leq 4\varphi_2(\delta). \tag{5.33}$$

Moreover, $\forall t \in \bar{t} \mapsto \bar{t}+1$,

$$w_\delta(t) = w_\delta(\bar{t}) + \int_{\bar{t}}^t w'_\delta(\eta)d\eta.$$

and

$$\begin{aligned} \|w_\delta(t)\|_{L^2} &\leq \|w_\delta(\bar{t})\|_{L^2} + \int_{\bar{t}}^t \|w'_\delta(\eta)\|d\eta \\ &\leq \|w_\delta(\bar{t})\|_{L^2} + \|w'_\delta(\bar{t})\|_{L^2(L^2)} \\ &\leq \|w_\delta(\bar{t})\|_{L^2} + \varphi_2(\delta). \end{aligned} \tag{5.34}$$

Setting $\tilde{Q}_\delta = t'_\delta \mapsto t''_\delta \times \Omega$, we obtain, by (1.22) and theorem V, since $p \geq 2$,

$$\begin{aligned} &\left|\int_{t'_\delta}^{t''_\delta} \langle \beta(x'(t+\delta)) - \beta(x'(t)), w_\delta(t)\rangle dt\right| \\ &= \left|\int_{\tilde{Q}_\delta} (\beta(x_t(t+\delta,\zeta)) - \beta(x_t(t,\zeta)))w_\delta(t,\zeta)dtd\zeta\right| \\ &\leq c_4 \int_{\tilde{Q}_\delta} (1+|x_t(t+\delta,\zeta)|^{p-2} + |x_t(t,\zeta)|^{p-2})\left|\frac{\partial w_\delta(t,\zeta)}{\partial t}\right| |w_\delta(t,\zeta)|dtd\zeta \\ &\leq c_4\left(\left\|\frac{\partial w_\delta}{\partial t}\right\|_{L^2(\tilde{Q}_\delta)} \|w_\delta\|_{L^2(\tilde{Q}_\delta)} + 2M_2^{p-2}\left\|\frac{\partial w_\delta}{\partial t}\right\|_{L^p(\tilde{Q}_\delta)} \|w_\delta\|_{L^p(\tilde{Q}_\delta)}\right). \end{aligned} \tag{5.35}$$

Recall that, by (1.21), the embedding of $H^1(\tilde{Q}_\delta)$ in $L^p(\tilde{Q}_\delta)$ is continuous, $\forall\delta$. Since $\frac{1}{2} \leq t''_\delta - t'_\delta \leq 1$, it follows from (5.35) (by (5.31) and (5.32)) that

$$\begin{aligned} &\left|\int_{t'_\delta}^{t''_\delta} \langle \beta(x'(t+\delta)) - \beta(x'(t)), w_\delta(t)\rangle dt\right| \\ &\leq \varphi_3(\delta)\|w_\delta\|_{L^p(\tilde{Q}_\delta)} \leq c_6\varphi_3(\delta)\|w_\delta\|_{H^1(\tilde{Q}_\delta)} \\ &= c_6\varphi_3(\delta)\left\{\int_{t'_\delta}^{t''_\delta} [\|w_\delta(t)\|^2_{H^1_0} + \|w'_\delta(t)\|^2_{L^2}]dt\right\}^{1/2} \\ &\leq c_6\varphi_3(\delta)\left\{\int_{t'_\delta}^{t''_\delta} \|w_\delta(t)\|^2_{H^1_0}dt\right\}^{1/2} + c_6\varphi_3(\delta)\|w'_\delta(\bar{t})\|_{L^2(L^2)} \\ &\leq \varphi_4(\delta)\left\{\int_{t'_\delta}^{t''_\delta} \|w_\delta(t)\|^2_{H^1_0}dt\right\}^{1/2} + \varphi_5(\delta). \end{aligned} \tag{5.36}$$

Let us now apply (5.26) to the function $w_\delta(t)$. Setting $t_1 = t'_\delta$, $t_2 = t''_\delta$, we obtain, by (5.22), (5.32), (5.33), (5.34), (5.36),

$$\int_{t'_\delta}^{t''_\delta} \|w_\delta(t)\|^2_{H^1_0} dt \le |(w'_\delta(t''_\delta), w_\delta(t''_\delta))_{L^2}| + |(w'_\delta(t'_\delta), w_\delta(t'_\delta))_{L^2}|$$

$$+ \int_{t'_\delta}^{t''_\delta} \{\|w'_\delta(t)\|^2_{L^2} + |\langle \beta(x'(t+\delta)) - \beta(x'(t)), w_\delta(t)\rangle|$$

$$+ |\langle f(t+\delta) - f(t), w_\delta(t)\rangle|\} dt$$

$$\le 2.4\varphi_2(\delta)(\|w_\delta(\bar t)\|_{L^2} + \varphi_2(\delta)) + \varphi_2^2(\delta) + \varphi_4(\delta) \tag{5.37}$$

$$\times \left\{\int_{t'_\delta}^{t''_\delta} \|w_\delta(t)\|^2_{H^1_0} dt\right\}^{1/2} + \varphi_5(\delta) + \varphi_0(\delta)\|w_\delta\|_{L^p(\hat Q_\delta)}$$

$$\le \varphi_6(\delta)\left\{\int_{t'_\delta}^{t''_\delta} \|w_\delta(t)\|^2_{H^1_0} dt\right\}^{1/2} + \varphi_7(\delta)\|w_\delta(\bar t)\|_{L^2} + \varphi_8(\delta),$$

where $\|w_\delta\|_{L^p(\hat Q_\delta)}$ is evaluated as in (5.36). It follows that

$$\int_{t'_\delta}^{t''_\delta} \|w_\delta(t)\|^2_{H^1_0} dt \le \varphi_6^2(\delta) + 2\varphi_7(\delta)\|w_\delta(\bar t)\|_{L^2} + 2\varphi_8(\delta). \tag{5.38}$$

Moreover, by (5.32) and (5.38),

$$\int_{t'_\delta}^{t''_\delta} \|w_\delta(t)\|^2_E dt = \int_{t'_\delta}^{t''_\delta} \{\|w_\delta(t)\|^2_{H^1_0} + \|w'_\delta(t)\|^2_{L^2}\} dt$$

$$\le \varphi_6^2(\delta) + 2\varphi_7(\delta)\|w_\delta(\bar t)\|_{L^2} + 2\varphi_8(\delta) + \varphi_2^2(\delta)$$

$$= 2\varphi_7(\delta)\|w_\delta(\bar t)\|_{L^2} + \varphi_9(\delta).$$

Hence there exists in $t'_\delta \vdash\!\dashv t''_\delta$ a point t^*_δ such that (since $t''_\delta - t'_\delta \ge \frac{1}{2}$)

$$\|w_\delta(t^*_\delta)\|^2_E \le 4\varphi_7(\delta)\|w_\delta(\bar t)\|_{L^2} + 2\varphi_9(\delta). \tag{5.39}$$

Let us apply relation (5.25) to $w_\delta(t)$, considering the interval $t^*_\delta \vdash\!\dashv \bar t + 1$; we obtain, by (5.22), (5.31), and (5.39),

$$\|w_\delta(\bar t+1)\|^2_E \le \|w_\delta(t^*_\delta)\|^2_E + 2\left\{\int_{t^*_\delta}^{\bar t+1} \|f(t+\delta) - f(t)\|^r_{L^r} dt\right\}^{1/r}$$

$$\times \left\{\int_{t^*_\delta}^{\bar t+1} \|w'_\delta(t)\|^p_{L^p} dt\right\}^{1/p}$$

$$\le 4\varphi_7(\delta)\|w_\delta(\bar t)\|_{L^2} + 2\varphi_9(\delta) + 2\varphi_0(\delta)\left(\frac{\varphi_0(\delta)}{c_3}\right)^{1/(p-1)}$$

$$= 4\varphi_7(\delta)\|w_\delta(\bar t)\|_{L^2} + \varphi_{10}(\delta).$$

Consequently, either (5.30) holds or (since $\|w_\delta(\bar t)\|_E \le \|w(\bar t+1)\|_E$)

$$\|w_\delta(\bar t+1)\|^2_E \le 4\varphi_7(\delta)\|w_\delta(\bar t+1)\|_E + \varphi_{10}(\delta). \tag{5.40}$$

Hence, by (5.40),

$$\|w_\delta(\bar t+1)\|^2_E \le 16\varphi_7^2(\delta) + 2\varphi_{10}(\delta) = \varphi_{11}^2(\delta). \tag{5.41}$$

From (5.17) and (5.28) it follows, in any case,

$$\|w_\delta(\bar{t}+1)\|_E \le \max(\|w_\delta(\bar{t})\|_E, \varphi_{11}(\delta)). \tag{5.42}$$

Let t be any point $\in J_0$. There exists a point $t_0 \in 0 \vdash\!\dashv 1$ and an integer $s \ge 0$ such that $t = t_0 + s$. Consider the points $t_j = t_0 + j$ ($j = 0, 1, 2, \cdots$) and observe that, by (5.42),

$$\|w_\delta(t_{j+1})\|_E \le \max(\|w_\delta(t_j)\|_E, \varphi_{11}(\delta)).$$

Thus, by (5.27),

$$\|w_\delta(t)\|_E = \|w_\delta(t_s)\|_E \le \max(\|w_\delta(t_0)\|_E, \varphi_{11}(\delta)) \le \max(\varphi_1(\delta), \varphi_{11}(\delta)), \tag{5.43}$$

that is, $x(t)$ is E-u.c. on J_0 and the theorem is proved.

The proof of theorem V is analogous to that of theorem VI, provided we substitute $x(t)$ for $w_\delta(t)$, relation (5.6) for (5.25), and suitable constants K_j for the functions $\varphi_j(\delta)$.

(c) *Proof of theorem VII.* Let us consider the sequence of functions $\{x_n(t)\}$ defined in the following way. When $t \ge -n$, $x_n(t)$ is the solution (which exists, is unique and E-continuous) satisfying the initial conditions $x_n(-n) = x'_n(-n) = 0$; when $t < -n$, it is $x_n(t) = 0$.

Moreover, if we set

$$f_n(t) = \begin{cases} f(t) & \text{when } t \ge -n, \\ 0 & \text{when } t < -n, \end{cases} \tag{5.44}$$

the function $x_n(t)$ is a solution on J corresponding to the known term $f_n(t)$.

We have, by theorem V,

$$\sup_J \|x_n(t)\|_E \le M_1, \qquad \sup_J \left\{ \int_0^1 \|x'_n(t+\eta)\|_{L^p}^p d\eta \right\}^{1/p} \le M_2. \tag{5.45}$$

Observe that, by the proof of theorem VI, we obtain, $\forall t \ge -n$, $\delta > 0$, $\|x_n(t+\delta) - x_n(t)\|_E \le \varphi(\delta)$, with $\varphi(\delta)$ independent of n and $\lim_{\delta \to 0} \varphi(\delta) = 0$.

Since $x_n(t) \underset{E}{=} 0$ for $t \le -n$, the same inequality holds on all J and the functions $x_n(t)$ are therefore E-equally u.c. on J.

Let us prove, first of all, that it is possible to select from $\{x_n(t)\}$ a subsequence (again denoted by $\{x_n(t)\}$) such that

$$\lim_{n \to \infty}{}^{*} x_n(t) \underset{E}{=} x(t) \qquad \forall t \in J. \tag{5.46}$$

The functions $(x_n(t), v)_E$ are, in fact, $\forall v \in E$, equally bounded and equally u.c. on J. As E is separable, we can assume that the sequence $\{(x_n(t), v)_E\}$ converges, $\forall v \in E$, uniformly on every bounded interval. Hence, (5.46) holds. Moreover,

$$\|x(t+\delta) - x(t)\|_E \le \min \lim_{n \to \infty} \|x_n(t+\delta) - x_n(t)\|_E \le \varphi(\delta). \tag{5.47}$$

By the first of (5.45), the functions $x_n(t)$ are, in particular, H_0^1-equally bounded and L_2-equally u.c. on J; since the embedding of H_0^1 in L^2 is compact, we can select, by the (vectorial) theorem of Ascoli-Arzelà, from $\{x_n(t)\}$ a subsequence (again denoted by $\{x_n(t)\}$) such that

(5.48) $$\lim_{n\to\infty} x_n(t) \underset{L^2}{=} x(t)$$

uniformly on every bounded interval.

Since the functions $x_n'(t)$ are L^2-equally u.c. on J, by (5.48) and a theorem of differentiation by series, we have

(5.49) $$\lim_{n\to\infty} x_n'(t) \underset{L^2}{=} x'(t),$$

uniformly on every bounded interval.

By the same procedure followed in theorem IV, it can be shown that, setting $Q_s = -s \vdash\!\dashv s \times \Omega$, we have, $\forall s = 1, 2, \cdots,$

(5.50) $$\lim_{n\to\infty}{}^{*}\ \beta\left(\frac{\partial x_n(t,\zeta)}{\partial t}\right) \underset{L^r(Q_s)}{=} \beta\left(\frac{\partial x(t,\zeta)}{\partial t}\right).$$

As $x_n(t)$ is a solution on J of the equation

(5.51) $$Ax_n(t) - x_n''(t) + f_n(t) = \beta(x_n'(t)),$$

it follows from (5.44), (5.45), (5.46), (5.50) that $x(t)$ is a bounded solution of (1.17); moreover, by (5.47), $x(t)$ is E-u.c. on J.

(*d*) *Proofs of theorems VIII, IX, and X.* Assume that there exist two solutions $x(t)$ and $y(t)$, such that (1.25) holds and

(5.52) $$\max \lim_{t\to-\infty} \|y(t)\|_E < +\infty.$$

Setting $w(t) = x(t) - y(t)$, we obtain, by (5.25),

(5.53) $$\|w(t_2)\|_E^2 - \|w(t_1)\|_E^2 + 2c_3 \int_{t_1}^{t_2} \|w'(t)\|_{L^p}^p dt \le 0.$$

The function $\|w(t)\|_E$ is therefore decreasing: since, by (1.25) and (5.52), $\sup_J \|w(t)\|_E < +\infty$, we have then

(5.54) $$\lim_{t\to-\infty} \|w(t)\|_E = N < +\infty.$$

The theorem will be proved if we can show that $N=0$, since, in that case, $\|w(t)\|_E \equiv 0$, that is, $x(t) \equiv y(t)$. For this we shall extend a procedure given by Amerio [1] for ordinary nonlinear systems.

By (5.53) and (5.54) we have (since the embedding of $L^p(L^p)$ $(p \ge 2)$ in $L^2(L^2)$ is continuous)

(5.55) $$\lim_{t\to-\infty} \|w'(t)\|_{L^p(L^p)} = \lim_{t\to-\infty} \|w'(t)\|_{L^2(L^2)} = 0.$$

Hence, by (5.54), (5.55),

$$\lim_{t\to-\infty} \|w(t)\|^2_{L^2(H_0^1)} = \lim_{t\to-\infty} (\|w(t)\|^2_{L^2(E)} - \|w'(t)\|^2_{L^2(L^2)}) = N^2. \tag{5.56}$$

On the other hand, analogously to (5.35) and (5.36), when $t \leq t_0$,

$$\begin{aligned} &\left| \int_0^1 \langle \beta(x'(t+\eta)) - \beta(y'(t+\eta)), w(t+\eta) \rangle d\eta \right| \\ &\leq c_4(\|w'(t)\|_{L^2(L^2)} \|w(t)\|_{L^2(L^2)} + 2M_0^{p-2} \|w'(t)\|_{L^p(L^p)} \|w(t)\|_{L^p(L^p)}) \\ &\leq c_7 \|w(t)\|_{L^2(E)} (\|w'(t)\|_{L^2(L^2)} + \|w'(t)\|_{L^p(L^p)}). \end{aligned} \tag{5.57}$$

Since $\|w(t)\|_{L^2(E)} \leq N$, it follows from (5.55), (5.57)

$$\lim_{t\to-\infty} \int_0^1 \langle \beta(x'(t+\eta)) - \beta(y'(t+\eta)), w(t+\eta) \rangle d\eta = 0. \tag{5.58}$$

Hence, considering (5.26) on the interval $t-1 \vdash\!\dashv t$, we obtain, by (5.56) and (5.58),

$$\lim_{t\to-\infty} [(w'(t-1), w(t-1))_{L^2} - (w'(t), w(t))_{L^2}] = N^2.$$

If $N > 0$, we would have, when $t \leq \bar{t}$,

$$(w'(t-1), w(t-1))_{L^2} \geq (w'(t), w(t))_{L^2} + \frac{N^2}{2}.$$

Hence, for $n = 1, 2, \cdots$,

$$(w'(\bar{t}-n), w(\bar{t}-n))_{L^2} \geq (w'(\bar{t}), w(\bar{t}))_{L^2} + n\frac{N^2}{2},$$

which is absurd, as $w(t)$ is E-bounded.

It must therefore be $N = 0$ and theorem VIII is proved.

The proof of theorem IX is similar to that of theorem VIII. By using the same procedure followed at § 4 one proves, finally, theorem X.

CHAPTER 8

RESULTS REGARDING OTHER FUNCTIONAL EQUATIONS

1. LINEAR FUNCTIONAL EQUATIONS

Favard's theory has been generalized (Amerio [15]), to the following second order functional equation

$$(1.1)\quad \int_J \{(x'(t),h'(t))_Y - (A(t)x(t),h(t)) + (B(t)x'(t),h(t))\}dt = \int_J (f(t),h(t))_Y dt.$$

In equation (1.1) we shall use the same notation as in Chapter 6, § 1 (*b*). For the general theory of (1.1), see Lions [2], Lions and Magenes [1].

The operators $A(t)$ and $B(t)$ are linear and bounded, $\forall t \in J$, from X to X and from Y to X, respectively; therefore

$$A(t) \in \mathscr{L}(X,X) = \mathscr{A}, \qquad B(t) \in \mathscr{L}(Y,X) = \mathscr{B}, \qquad \forall t \in J.$$

$\mathscr{A}$ and $\mathscr{B}$ are Banach spaces and we shall assume that $A(t)$ and $B(t)$ are *continuous functions*, on J, in their respective uniform topologies, that is, as functions with values in $\mathscr{A}$ and $\mathscr{B}$.

In (1.1), the *unknown function* $x(t)$, the *test function* $h(t)$ and the *known term* $f(t)$ belong to the following functional spaces:

$$(1.2)\quad \begin{aligned} x(t),h(t) &\in L^2_{\text{loc}}(J;X); \\ x'(t),h'(t),f(t) &\in L^2_{\text{loc}}(J;Y). \end{aligned}$$

Moreover, $h(t)$ *has compact support and* (1.1) *must hold for all test functions* $h(t)$.

We shall denote by W the space of functions $g=\{g(\eta);\ \eta \in 0 \vdash\!\dashv 1\}$ such that $g(\eta) \in L^2(0 \vdash\!\dashv 1;\ X) = L^2(X)$, $g'(\eta) \in L^2(0 \vdash\!\dashv 1;\ Y) = L^2(Y)$.

Hence, if $z(t)=\{z(t+\eta);\ \eta \in 0 \vdash\!\dashv 1\}$ is such that $z(t) \in L^2_{\text{loc}}(J;X)$, $z'(t) \in L^2_{\text{loc}}(J;Y)$, we have

$$(1.3)\quad \begin{aligned} \|z(t)\|_W &= \left\{\int_0^1 (\|z(t+\eta)\|^2 + \|z'(t+\eta)\|_Y^2)d\eta\right\}^{1/2} \\ &= \{\|z(t)\|^2_{L^2(X)} + \|z'(t)\|^2_{L^2(Y)}\}^{1/2}. \end{aligned}$$

Thus $z(t)$ is W-continuous. Moreover, one proves easily that $z'(t)$ defines also the $L^2(Y)$-derivative of $z(t)$, that is,

$$\lim_{\tau\to 0}\left\|\frac{z(t+\tau)-z(t)}{\tau}-z'(t)\right\|_{L^2(Y)} = 0.$$

Let now $z(t)=\{z(t+\eta);\ \eta\in 0\vdash\!\dashv 1\}$ be a W-bounded function. We shall set

$$\varphi(z,\tau) = \sup_J \|z(t+\tau)-z(t)\|_W \qquad (\forall\tau\in J), \tag{1.4}$$

$$\mu(z) = \sup_J \|z(t)\|_W. \tag{1.5}$$

Let Γ_z be the set of W-bounded solutions $x(t)$ such that

$$\varphi(x;\ \tau) \leq \varphi(z;\ \tau) \qquad \forall\tau\in J. \tag{1.6}$$

Moreover, let Λ_z be the set of those eigensolutions $u(t)$ of the homogeneous equation

$$\int_J \{(u'(t),h'(t))_Y-(A(t)u(t),h(t))+(B(t)u'(t),h(t))\}dt = 0 \tag{1.7}$$

which are differences between functions $\in\Gamma_z$.

Let us state now the *minimax theorem.*

I. *Assume that:*

(1) *There exists a W-bounded function $z(t)$ such that the set Γ_z is not empty;*

(2) $\forall u(t)\in\Lambda_z$,

$$\inf_J \|u(t)\|_W > 0. \tag{1.8}$$

Then, if

$$\tilde{\mu} = \inf_{\Gamma_z} \mu(x),$$

there exists, in Γ_z, one, and only one, solution, $\tilde{x}(t)$, such that

$$\mu(\tilde{x}) = \tilde{\mu}.$$

It can be observed, comparing this theorem with the minimax theorem regarding the wave equation and with the one of Favard for ordinary linear systems, that the class of solutions Γ_z is here more restricted, owing to condition (1.6) (the same observation holds for the Schrödinger type equation).

On the other hand, the rather restrictive condition (1.8) is here broadened, because we impose it only on the functions $u(t)\in\Lambda_z$ and not on all the W-bounded eigensolutions.

Assume, from now on, *that the operators* $A(t)$, $B(t)$ *are, respectively,* $\mathscr{A}$*-a.p. and* $\mathscr{B}$*-a.p. and that* $f(t)$ *is* Y-S^2 *w.a.p.*

If $s=\{s_n\}$ is a *regular* sequence with respect to $A(t)$, $B(t)$, $f(t)$ and if $\mathscr{S}$ is the family of such sequences, we obtain, uniformly,

$$\lim_{n\to\infty} A(t+s_n) = A_s(t),$$

$$\lim_{n\to\infty} B(t+s_n) = B_s(t), \tag{1.9}$$

$$\lim_{n\to\infty}{}^{*} f(t+s_n) = f_s(t),$$

with $A_s(t)$ $\mathscr{A}$-a.p., $B_s(t)$ $\mathscr{B}$-a.p., $f_s(t)$ Y-S^2 w.a.p.

We can, therefore, consider, $\forall s \in \mathscr{S}$, the equation

$$\int_J \{(x'(t),h'(t))_Y-(A_s(t)x(t),h(t))+(B_s(t)x'(t),h(t))\}dt = \int_J (f_s(t),h(t))_Y dt, \tag{1.10}$$

and the corresponding sets $\Gamma_{z,s}$, $\Lambda_{z,s}$.

It can be shown that, *if* (1.1) *admits a* W*-bounded solution* $z(t)$, *then each equation* (1.10) *has one* W*-bounded solution*, $z_s(t)$, *such that*

$$\varphi(z_s;\ \tau) \leq \varphi(z;\ \tau).$$

Hence the set $\Gamma_{z,s}$ is not empty, $\forall s \in \mathscr{S}$.

We now state two almost-periodicity theorems.

II. *Assume that:*

(1) *Equation* (1.1) *has one* W*-bounded solution*, $z(t)$;

(2) $\forall s \in \mathscr{S}$ *and* $u(t) \in \Lambda_{z,s}$,

$$\inf_J \|u(t)\|_W > 0. \tag{1.11}$$

Then the minimal solution $\tilde{x}(t)$ (which, by theorem I, exists and is unique), *is* W*-w.a.p.*

Observe that $z(t)$ is supposed to be only *bounded* (as in Favard's theory) and not *bounded* and *w.u.c.* (as in theorem V, Chapter 6, for the Schrödinger-type equation).

This follows from the fact that here we consider (by the definition itself of the space W) the X- or Y-almost-periodicity in the sense of Stepanov.

III. *Assume that the hypotheses of theorem II hold, with* $z(t)$ W*-u.c.; assume, moreover, that the embedding of* X *in* Y *is compact and that the operator* $A(t)$ *satisfies the ellipticity condition*

$$\mathscr{R}(A(t)x,x) \geq \nu\|x\|^2 \qquad \forall x \in X \qquad (\nu > 0). \tag{1.12}$$

Then the minimal solution, $\tilde{x}(t)$, *is* W*-a.p.*

It may be observed that, for the problem of the vibrating membrane treated at Chapter 5, condition (1.12) is obviously verified, as is the hypothesis that the embedding of X in Y is compact, being, in this case, $X = H_0^1(\Omega)$, $Y = L^2(\Omega)$, where the set Ω is *bounded*. It would be interesting to prove theorem III, reducing the hypothesis that $z(t)$ is W-u.c.; see, for this, Prouse [5], Ricci and Vaghi [1]. See also, for various results, Günzler and Zaidman [1], [2], Fojas and Zaidman [1], Likov [1].

OBSERVATION I. We shall say that $x(t)$ is *a solution of* equation (1.1) *on the interval* $J^+ = \bar{t}\,\overline{}\,+\infty$ if (1.1) is satisfied for all test functions $h(t)$ with support $\subset J^+$.

Assume now that $A(t)$, $B(t)$, *and* $f(t)$ *are, respectively,* $\mathscr{A}$*-a.p.*, $\mathscr{B}$*-a.p.*, Y-S^2 *w.a.p.*

One proves then that, *if equation* (1.1) *admits, on the interval* J^+, *a bounded solution* $\bar{z}(t)$, *there exists also a solution,* $z(t)$, *which is bounded on all* J.

It is, by assumption,

$$\sup_{J^+} \|\bar{z}(t)\| = \bar{M} < +\infty. \tag{1.13}$$

Consider now the sequence $\{\bar{z}(t+n)\}$, $n=1, 2, \cdots$; the function $\bar{z}(t+n)$ satisfies on the interval $\bar{t}-n\,\overline{}\,+\infty$ the equation corresponding to the operators and known term $A(t+n)$, $B(t+n)$, $f(t+n)$.

By the almost-periodicity of $A(t)$, $B(t)$, $f(t)$, and by (1.13), we can select a sequence $\{s_n\}$ of positive integers such that relations (1.9) hold and we have, on every interval $\Delta_p = -p \vdash\!\dashv p$ $(p=1, 2, \cdots)$,

$$\lim_{n\to\infty}{}^{*}\ \bar{z}(t+s_n) \underset{L^2(\Delta_p;X)}{=} z_s(t),$$

$$\lim_{n\to\infty}{}^{*}\ \bar{z}'(t+s_n) \underset{L^2(\Delta_p;Y)}{=} z_s'(t),$$

that is, $\forall t \in J$, by (1.3),

$$\lim_{n\to\infty}{}^{*}\ \bar{z}(t+s_n) \underset{W}{=} z_s(t).$$

It follows, by (1.13), that

$$\sup_{J} \|z_s(t)\|_W \leq \bar{M} \tag{1.14}$$

and $z_s(t)$ will be a solution, on all J, of equation (1.10).

We select now, as before, from the sequence $-s = \{-s_n\}$, a sequence $-s' = \{-s_n'\}$ such that, $\forall t \in J$,

$$\lim_{n\to\infty}{}^{*}\ z_s(t-s_n') \underset{W}{=} z(t),$$

which implies

$$\sup_{J} \|z(t)\| \leq \bar{M}.$$

Since, by (1.9) and by the almost-periodicity,

$$\lim_{n\to\infty} A_s(t-s'_n) = A(t),$$

$$\lim_{n\to\infty} B_s(t-s'_n) = B(t),$$

$$\lim_{n\to\infty}{}^{*} f_s(t-s'_n) = f(t),$$

uniformly, $z(t)$ is a W-bounded solution, on all J, of the equation (1.1).

Observe that if $\bar{z}(t)$, W-bounded on J^+, is also W-u.c. (or W-w.u.c.) on J^+, then the same properties hold, on all J, for $z(t)$.

The above procedure holds obviously for other equations (in particular, for those considered in Chapters 5 and 6).

Observation II. It is possible to effect the *harmonic analysis of the minimal solution* $\tilde{x}(t)$.

Observe in fact that, because of the *uniqueness* of the minimal solution, every sequence $s \in \mathscr{S}$ results also regular with respect to $\tilde{x}(t)$. From theorem X of Chapter 2, it follows that *every characteristic exponent of* $\tilde{x}(t)$ *is a linear combination, with integer coefficients, of a finite number of characteristic exponents of* $A(t)$, $B(t)$, $f(t)$.

Also this observation can obviously be applied to other equations.

Observation III. Theorems I, II, and III hold also for equations in more general spaces (Amerio [7]); more precisely, if we assume that X is *uniformly convex* and Y *semicomplete*. In this case, the proof of theorem III can be obtained by assuming that a continuous dependence theorem from the operator and the known term holds; such a theorem has, however, been proved only in particular cases. Zaidman [2] has shown that this is true for the *wave equation*—more precisely, for the *Cauchy problem* under the assumption that the initial values are given on the whole of R^m (that is, $\Omega = R^m$) and $a_0(\zeta) \geq \rho > 0$. In this case, *the E-bounded eigensolutions are not,in general, a.p. However, the minimal solution,* $\tilde{x}(t)$ *is a.p. if* $f(t)$ *is* L^2*-a.p.* See also the results of Zikov [1], [2].

Observation IV. We wish to outline a method, different from the ones given so far, used by Zaidman [7] to prove the almost-periodicity of the bounded solutions of a second-order linear functional equation (for a similar application to other equations, see Zaidman [5], [6], [8]).

Let X and Y be two separable Hilbert spaces, with $X \subseteq Y$ and dense in Y, the embedding of X in Y being continuous. Moreover, let $\gamma(u,v)$ be a sesquilinear form, continuous on $X \times X$ such that

$$\gamma(u,v) = \overline{\gamma(v,u)}, \qquad \forall u, v \in X,$$

and, $\forall \beta > 0$, there exists $\alpha > 0$ such that

$$\gamma(u,u) + \beta\|u\|_Y^2 \geq \alpha\|u\|_X^2. \tag{1.15}$$

The form $\gamma(u,v)$ defines then a linear, positive, self-adjoint, unbounded operator Γ, of domain $D(\Gamma)$, by the relation

$$(1.16) \qquad \gamma(u,v) = (\Gamma u,v)_Y, \qquad \forall u \in D(\Gamma).$$

Consider the equation

$$(1.17) \qquad x''(t) - \Gamma x(t) = f(t),$$

which obviously generalizes the Poisson equation (the boundary condition being contained in the definition itself of the spaces X and Y).

Let us assume that $f(t) \in L^2_{\text{loc}}(J;Y)$: *then there exists at least one solution,* $x(t)$, *on* J, *of* (1.17). By this we mean that *there exists a function* $x(t) \in C^0(J;X)$, *with* $x'(t) \in L^2_{\text{loc}}(J;X) \cap C^0(J;Y)$, $x''(t) \in L^2_{\text{loc}}(J;Y)$, $\Gamma x(t) \in L^2_{\text{loc}}(J;Y)$, *satisfying equation* (1.17) *a.e. on* J. Moreover, *if* $f(t)$ *is* Y*-continuous and bounded, then any* Y*-bounded solution* $x(t)$ *is* X*-bounded and the derivative* $x'(t)$ *is* Y*-bounded.*

We shall now give the main points of Zaidman's proof of the following almost-periodicity theorem.

Let $f(t)$ *be* Y*-a.p. and* $x(t)$ *be a solution on* J *of* (1.17). *Then, if* $x(t)$ *is* Y*-bounded on* J ($\sup_J \|x(t)\|_Y = K_1 < +\infty$), $x(t)$ *is* X*-a.p. and* $x'(t)$ *is* Y*-a.p.*

Observe at first that, by (1.15), (1.16), we have, $\forall u \in D(\Gamma)$,

$$(\Gamma u,u)_Y \geq 0.$$

Since Γ is self-adjoint, we can set

$$\Gamma u = \int_0^\infty \sigma dE_\sigma u,$$

E_σ being the spectral family associated to Γ, normalized by the relation $E_{\sigma^-} = E_\sigma$.

Setting $E_{p,q} = E_q - E_p$, the following statements hold.

(i_1) *If* $0 \leq p \leq q < +\infty$, *then* $E_{p,q} Y \subset D(\Gamma)$ *and we have*

$$\|\Gamma E_{p,q} u\|_Y \leq q\|u\|_Y.$$

(i_2) *If* $x(t) \in D(\Gamma)$ *a.e. on* J, $x''(t) \in L^2_{\text{loc}}(J;Y)$ *and equation* (1.17) *is satisfied a.e. on* J, *then* $\forall p$, q (*with* $0 \leq p < q < +\infty$) $E_{p,q}x(t) \in C^2(J;Y)$ *and*

$$\frac{d^2}{dt^2} E_{p,q}x(t) = \Gamma E_{p,q}x(t) + E_{p,q}f(t) \qquad \forall t \in J;$$

moreover, $\|E_{p,q}x(t)\|_Y \leq (1/p)\|E_{p,q}f(t)\|_Y$ *and* $E_{p,q}x(t)$ *is* Y*-a.p.*

Let us now give the proof of the Y-almost-periodicity of $x(t)$. Observing that $E_0 = 0$ and $E_\sigma \to I$ when $\sigma \to +\infty$, we set

$$I = E_{0,0^+} + E_{0^+,1} + E_{1,+\infty} = E_{0,0^+} + E_{0^+,1} + \sum_1^\infty{}_n E_{n,n+1}.$$

It will therefore be sufficient to prove that the functions $E_{0,0^+}x(t)$, $E_{0^+,1}x(t)$, $E_{1,+\infty}x(t)$ are Y-a.p.

Observe, first, that

$$\frac{d^2}{dt^2} E_{0,0^+}x(t) = E_{0,0^+}x''(t) = E_{0,0^+}f(t).$$

Since $E_{0,0^+}x'(t)$ is Y-bounded and $E_{0,0^+}f(t)$ is Y-a.p., the function $E_{0,0^+}x'(t)$ is, by theorem II, Chapter 4, Y-a.p.; also $E_{0,0^+}x(t)$ is then Y-a.p.

Consider now the function $E_{0^+,1}x(t)$ and decompose $E_{0^+,1}$ in the following way

$$E_{0^+,1} = \sum_{1}^{\infty}{}_n E_{1/(n+1),1/n} = \sum_{1}^{\infty}{}_n F_n.$$

Since $F_n^2 = F_n$ and $F_n F_m = 0$ for $m \neq n$, we obtain, by (i_2),

$$\frac{d^2}{dt^2}(F_n x(t)) = \Gamma F_n(F_n x(t)) + F_n f(t) \tag{1.18}$$

and $F_n x(t)$ is Y-a.p., $\forall n = 1, 2, \cdots$.

Moreover,

$$E_{0^+,1}x(t) \underset{Y}{=} \sum_{1}^{\infty}{}_n F_n x(t), \qquad \forall t \in J.$$

Observe now that (as the range of the Y-a.p. function $E_{0^+,1}f(t)$ is r.c. in Y) the series $\sum_{(1)}^{(\infty)}{}_n F_n f(t)$ Y-converges to $E_{0^+,1}f(t)$ uniformly on J.

Also the series $\sum_{(1)}^{(\infty)}{}_n \Gamma F_n F_n x(t)$ Y-converges uniformly on J. In fact, by (i_1),

$$\|\Gamma F_n(F_n x(t))\|_Y \leq \frac{1}{n} \sup_J \|x(t)\|_Y = \frac{K_1}{n}$$

and, consequently (since the elements in the series are mutually orthogonal),

$$\left\| \sum_{1}^{\infty}{}_n \Gamma F_n F_n x(t) \right\|_Y^2 \leq K_1^2 \sum_{1}^{\infty}{}_n \frac{1}{n^2}.$$

Hence, by (1.18), the series

$$\sum_{1}^{\infty}{}_n \frac{d^2}{dt^2} F_n x(t)$$

converges uniformly on J to a Y-a.p. function.

As the function $x'(t)$ is Y-bounded on J, $\|\sum_{(1)}^{(\infty)}{}_n (d/dt)F_n x(t)\|_Y \leq K_2 < +\infty$.

By theorem II, Chapter 4, the function $\sum_{(1)}^{(\infty)}{}_n (d/dt)F_n x(t) = (d/dt)E_{0^+,1}x(t)$ is then Y-a.p. By the same theorem, $E_{0^+,1}x(t)$, being Y-bounded, is therefore Y-a.p.

Let us finally consider the term $E_{1,+\infty}x(t) = \sum_{(1)}^{(\infty)}{}_n E_{n,n+1}x(t)$. By (i_2), each element of the series is Y-a.p. and

$$\|E_{n,n+1}x(t)\|_Y \leq \frac{1}{n} \sup_J \|f(t)\|_Y. \tag{1.19}$$

Bearing in mind that its terms are mutually orthogonal, the series $\sum_{(1)}^{(\infty)}{}_n E_{n,n+1}x(t)$ is, by (1.19), Y-uniformly convergent on J and its sum is therefore Y-a.p.

Let τ be an ε-almost-period of $x(t)$ and $f(t)$. The function $x(t+\tau)-x(t)$ is obviously a solution of the equation

$$\frac{d^2}{dt^2}(x(t+\tau)-x(t)) = \Gamma(x(t+\tau)-x(t))+f(t+\tau)-f(t)$$

and one proves the relation

$$\sup_J \|x(t+\tau)-x(t)\|_X = \varepsilon K_3,$$

K_3 being a suitable positive constant; therefore, $x(t)$ is X-a.p.

By an analogous procedure it can be shown that $x'(t)$ is Y-a.p.

2. A NONLINEAR HEAT EQUATION

Let us consider the following nonlinear heat equation, in one dimension:

$$x_t = x_{\zeta\zeta} - f(t,\zeta,x,x_\zeta) \tag{2.1}$$

where $x(t,\zeta)$ is the unknown function and $f(t,\zeta,x,p)$ is a given function of the variables t, ζ, x, p, continuous on the set

$$\zeta \in 0 \vdash\!\dashv l (l < +\infty);\ t,x,p \in J.$$

All functions are supposed to be *real.*

We shall say, moreover, that *$f(t,\zeta,x,p)$ is a.p. with respect to t if, $\forall M > 0$, the function*

$$F_M(t) = \{f(t,\zeta,x,p);\ 0 \le \zeta \le l,\ |x| \le M,\ |p| \le M\}$$

is a.p. from J to the Banach space C_M^0, of functions $\varphi(\zeta,x,p)$ continuous on the domain $0 \le \zeta \le l$, $|x| \le M$, $|p| \le M$.

Consider now the *classical solutions* $x(t,\zeta)$ of (2.1), continuous on the strip $J \times 0 \vdash\!\dashv l$, together with their derivatives $x_\zeta(t,\zeta)$, $x_{\zeta\zeta}(t,\zeta)$ (which implies, by (2.1), that $x_t(t,\zeta)$ is also continuous), and satisfying the *boundary condition*

$$x(t,0) = x(t,l) = 0 \qquad (t \in J). \tag{2.2}$$

Let $C^2 = C^2(0 \vdash\!\dashv l)$ be the Banach space of all functions $g = g(\zeta)$, continuous on $0 \vdash\!\dashv l$, together with their derivatives $g'(\zeta)$ and $g''(\zeta)$.

Setting $x(t) = \{x(t,\zeta);\ \zeta \in 0 \vdash\!\dashv l)$, the *solution* $x(t)$, $t \in J$ is *C^2-continuous*, and the *derivative* $x'(t) = \{x_t(t,\zeta);\ \zeta \in 0 \vdash\!\dashv l\}$ is *C^0-continuous.*

The problem of the *existence of a C^2-a.p. solution* $x(t)$ (when $f(t,\zeta,x,p)$ is a.p. with respect to t) is connected, as we have previously observed, with the problem of the *existence and uniqueness of a C^2-bounded solution.* On this subject the following statements hold (Vaghi [4]).

I (*Existence and uniqueness theorem*). *Assume that:*

(1) $f(t,\zeta,x,p)$ *satisfies*, $\forall M>0$, *a Hölder condition with exponent* λ, $0<\lambda<1$, *on the domain* $(t\in J;\, 0\le\zeta\le l;\, |x|,\, |p|\le M)$;

(2) *We have* ($\forall\zeta\in 0\vdash\!\dashv l$ *and* $\forall t, x, p\in J$)

$$|f(t,\zeta,x,p)| \le \alpha(|z|)+\beta(|z|)\gamma(|p|), \tag{2.3}$$

where $\alpha(\rho)$, $\beta(\rho)$, $\gamma(\rho)$ *are positive and nondecreasing functions on the interval* $0\le\rho<+\infty$, *and*

$$\lim_{\rho\to\infty}\frac{\gamma(\rho)}{\rho^2} = 0; \tag{2.4}$$

(3) $f(t,\zeta,x,p)$ *is an increasing function of* x—*precisely*:

$$w\{f(t,\,\zeta,\,x+w,\,p)-f(t,\zeta,x,p)\} \ge kw^2,\ (k = const>0). \tag{2.5}$$

Then problems (2.1) *and* (2.2) *admit one and only one C^2-bounded solution* $x(t)$.

II (*Almost-periodicity theorem*). *Assume that the hypotheses* (1), (2), (3) *of theorem I are satisfied; assume, moreover, that* $f(t,\zeta,x,p)$ *is a.p. with respect to* t.

Then the C^2-bounded solution $x(t)$ (which exists and is unique by theorem I) *is also C^2-a.p.*

Hence, by (2.1), *the derivative* $x'(t)$ *is C^0-a.p.*

OBSERVATION V. The proof of theorem I is related to that previously given by Prodi [1] for proving the existence of one *periodic solution, at least*, of problem (2.1) and (2.2), under the hypothesis that $f(t,\zeta,x,p)$ is a *periodic* function of t; in particular, hypothesis (2.4) has already been assumed by Prodi.

A condition like (2.5) has been set by Vaghi [3] for proving the *uniqueness* of the *bounded* solutions, for the problem analogous to (2.1) and (2.2), in m-dimensional space. For an extension of theorems I, II to the equation

$$x_t = x_{\zeta\zeta}-f(t,\zeta,x,x_\zeta)+\varphi(t,\zeta,x)x_\zeta^2$$

see Vaghi [5].

3. THE NAVIER-STOKES EQUATION

(*a*) Let $\Omega\subset R^m(\zeta=(\zeta_1,\cdots,\zeta_m))$ be a bounded, open, and connected set, satisfying the cone property (see Chapter 7, § 1).

We shall denote by J_0 the interval $0\vdash\!\!-+\infty$ and set $J_0^+=0\!-\!\!-+\infty$, $Q=J_0\times\Omega$, $Q^+=J_0^+\times\Omega$.

Let $x(t,\zeta)$ be a real vector with components $\theta_1(t,\zeta),\cdots,\theta_m(t,\zeta)$ and $|x(t,\zeta)|=\{\sum_{(1)}^{(m)}{}_j\,\theta_j^2(t,\zeta)\}^{1/2}$ its Euclidean norm; let $p(t,\zeta)$ be a scalar quantity and $f(t,\zeta)$ a second real vector, with components $\gamma_1(t,\zeta),\cdots,\gamma_m(t,\zeta)$. Denote, finally, by Δ_ζ and div_ζ, respectively, the Laplace and divergence operators with respect to the "space" variables:

$$\Delta_\zeta\theta_j(t,\zeta) = \sum_1^m{}_k \frac{\partial^2\theta_j(t,\zeta)}{\partial\zeta_k^2}, \qquad \mathrm{div}_\zeta\, x(t,\zeta) = \sum_1^m{}_j \frac{\partial\theta_j(t,\zeta)}{\partial\zeta_j}.$$

The Navier-Stokes system, in its classical formulation, is

$$\text{(3.1)}\qquad \begin{cases} \dfrac{\partial\theta_j}{\partial t} - \mu\Delta_\zeta\theta_j + \displaystyle\sum_1^m{}_k \frac{\partial\theta_j}{\partial\zeta_k}\,\theta_k = -\frac{\partial p}{\partial\zeta_j} + \gamma_j \qquad (j=1,\cdots,m) \\ \mathrm{div}_\zeta\, x = 0. \end{cases}$$

This system describes, as is well known, the motion of an *incompressible fluid*, of unit density, with *viscosity coefficient* μ, subject to the *external force* $f(t,\zeta)$; the unknown functions $x(t,\zeta)$ and $p(t,\zeta)$ represent, respectively, the *velocity* of the fluid and its *pressure*.

We shall assume that $x(t,\zeta)$ satisfies the boundary condition

$$\text{(3.2)}\qquad x(t,\zeta)|_{\zeta\in\partial\Omega} = 0 \qquad (t\in J_0).$$

Let us give to the problem considered above a weak formulation, due to Hopf [1], to which we shall always refer in this section.

Let $h(t,\zeta)$ be a vector with components $\sigma_1(t,\zeta),\cdots,\sigma_m(t,\zeta)$, indefinitely differentiable and with compact support on the cylinder Q^+; moreover, let $\mathrm{div}_\zeta\, h(t,\zeta)=0$.

Applying Green's formula in an obvious way, it follows from (3.1) that

$$\text{(3.3)}\qquad \int_Q \left\{\sum_1^m{}_j \frac{\partial\theta_j}{\partial t}\sigma_j - \mu\sum_1^m{}_j \sigma_j\Delta_\zeta\theta_j + \sum_{j,k}^{1\ldots m} \sigma_j\frac{\partial\theta_j}{\partial\zeta_k}\theta_k\right\}dtd\zeta$$

$$= \int_Q \left\{-\sum_1^m{}_j \theta_j\frac{\partial\sigma_j}{\partial t} + \mu\sum_{j,k}^{1\ldots m}\frac{\partial\sigma_j}{\partial\zeta_k}\frac{\partial\theta_j}{\partial\zeta_k} + \sum_{j,k}^{1\ldots m}\sigma_j\frac{\partial\theta_j}{\partial\zeta_k}\theta_k\right\}dtd\zeta$$

$$= \int_Q \left\{-\sum_1^m{}_j \sigma_j\frac{\partial p}{\partial\zeta_j} + \sum_1^m{}_j \sigma_j\gamma_j\right\}dtd\zeta = \int_Q \sum_1^m{}_j \sigma_j\gamma_j dtd\zeta,$$

since

$$\int_Q \sum_1^m{}_j \sigma_j\frac{\partial p}{\partial\zeta_j}dtd\zeta = -\int_Q p\,\mathrm{div}_\zeta\, h dtd\zeta = 0.$$

Let us now define the following functional spaces.

Let $\mathcal{N}$ be the linear manifold of m-component vectors $u=u(\zeta)$ indefinitely differentiable on Ω, with null divergence and compact support on Ω and let N and N^1 be the closures of $\mathcal{N}$ in L^2 and H_0^1, respectively.

Observe that N and N^1 are subspaces of L^2 and H_0^1, respectively. We can therefore set

$$(3.4)\qquad \begin{aligned} (x_1,x_2)_N &= (x_1,x_2)_{L^2} = \int_\Omega \sum_1^m{}_j\, \theta_{1j}(\zeta)\theta_{2j}(\zeta)d\zeta,\\ (x_1,x_2)_{N^1} &= (x_1,x_2)_{H_0^1} = \int_\Omega \sum_{j,k}^{1\dots m} \frac{\partial\theta_{1j}(\zeta)}{\partial\zeta_k}\frac{\partial\theta_{2j}(\zeta)}{\partial\zeta_k}\,d\zeta. \end{aligned}$$

We shall, moreover, set

$$(3.5)\qquad b(x_1,x_2,x_3) = \int_\Omega \sum_{j,k}^{1\dots m} \theta_{3j}(\zeta)\frac{\partial\theta_{2j}(\zeta)}{\partial\zeta_k}\,\theta_{1k}(\zeta)d\zeta.$$

By (3.4) and (3.5), equation (3.3) can be written (setting, as usual, $x(t)=\{x(t,\zeta);\ \zeta\in\Omega\}$, etc.)

$$(3.6)\qquad \int_{J_0}\{\mu(x(t),h(t))_{H_0^1} - (x(t),h'(t))_{L^2} + b(x(t),x(t),h(t)) - (f(t),h(t))_{L^2}\}dt = 0.$$

Equation (3.6) is the *weak form* of system (3.1) and will be called the *Navier-Stokes equation*.

We shall simply state the results concerning (3.6) which are most significant in the frame of this book; for the proofs, we shall refer to the original papers. For the general theory of the Navier-Stokes equation, see Ladyženskaja [3].

Observe that, if $x(t)$ is a solution of (3.6) and if the corresponding function $x(t,\zeta)$ is sufficiently smooth, then $x(t,\zeta)$ satisfies the equation

$$\int_Q \sum_1^m{}_j\, \sigma_j(t,\zeta)\left\{\frac{\partial\theta_j}{\partial t} - \mu\Delta_\zeta\theta_j + \sum_1^m{}_k\,\frac{\partial\theta_j}{\partial\zeta_k}\,\theta_k - \gamma_j\right\}dtd\zeta = 0,$$

$\forall h(t,\zeta)$ of the class defined above. Hence, setting $\sigma_j(t,\zeta)=\sigma_j(\zeta)\varphi_j(t)$, with $\varphi_j(t)\in\mathscr{D}(J_0^+)$, we obtain

$$(3.7)\qquad \int_\Omega \sum_1^m{}_j\, \sigma_j(\zeta)\left\{\frac{\partial\theta_j}{\partial t} - \mu\Delta_\zeta\theta_j + \sum_1^m{}_k\,\frac{\partial\theta_j}{\partial\zeta_k}\,\theta_k - \gamma_j\right\}d\zeta = 0,$$

$\forall h(\zeta)=\{\sigma_1(\zeta),\cdots,\sigma_m(\zeta)\}\in\mathscr{D}(\Omega)$, with $\mathrm{div}_\zeta\, h(\zeta)=0$.

Relation (3.7) implies (see Prodi [4]) that there exists a function $p(t,\zeta)$ such that

$$\frac{\partial\theta_j}{\partial t} - \mu\Delta_\zeta\theta_j + \sum_1^m{}_k\,\frac{\partial\theta_j}{\partial\zeta_k}\,\theta_k - \gamma_j = -\frac{\partial p}{\partial\zeta_j}.$$

Hence, system (3.1) is verified.

When the functions considered are sufficiently smooth, the classical and the weak formulations are therefore equivalent.

(*b*) Let us, from now on, assume that the *unknown function* $x(t)$, the *known term* $f(t)$, and the *test function* $h(t)$, in (3.6), satisfy the following conditions:

(i_1) $x(t) \in L^1_{\text{loc}}(J_0;N^1) \cap L^\infty(J_0;N)$;

(i_2) $h(t) \in C^0(J_0;N^1)$, *has compact support on* J_0^+ *and is such that* $h'(t) \in L^1_{\text{loc}}(J_0;N)$;

(i_3) $f(t) \in L^1_{\text{loc}}(J_0;L^2)$.

Then $x(t)$ is a *solution* on J_0 of (3.6) if it satisfies that equation $\forall h(t)$ verifying (i_2).

Other weak formulations, slightly different from the one given above, are due to Leray [1] and to Kieselev and Ladyženskaja [1].

(*c*) We now state the fundamental theorem of Hopf [1] regarding the *existence* of a solution of the initial value problem.

I. $\forall x_0 \in N$, *there exists on* J_0 *at least one solution of* (3.6) *which is* N-*w.c. on* J_0 *and satisfies the initial condition*

$$x(0) = x_0. \tag{3.8}$$

It has not, so far, been possible to obtain the corresponding uniqueness theorem. *Uniqueness has been proved in the case* $m=2$ (Lions and Prodi [1]), utilizing an inequality due to Ladyženskaja [2]:

$$\|x\|_{L^4} \leq \sqrt{2}\, \|x\|_{L^2} \|x\|_{H_0^1},$$

which holds only if Ω is a two dimensional set.

II. *If* $m=2$, *the solution* $x(t)$ *of the initial value problem is unique.*

We also recall, in the case $m=3$, the following result, due to Prodi [3].

III. *Assume that* $m=3$ *and let* $x(t)$ *be a solution of the initial value problem. If there exists* $p>3$ *such that*

$$x(t) \in L^{2p/(p-3)}_{\text{loc}}(J_0;L^p),$$

then the same problem does not admit any other solution.

We now give some results concerning the regularization and other properties of the solutions.

IV (*Prodi* [3], *Lions and Prodi* [1]). *Let* $m=2$. *Then,* $\forall$ *solution* $x(t)$, *the following properties hold*:

(1) *After an eventual modification on a set of measure zero,* $x(t)$ *is* N-*continuous*;

(2) *Denoting by* $v(t)$ *an arbitrary* N^1*-continuous function, such that* $v'(t) \in L^2_{\mathrm{loc}}(J_0;N)$, *we have*, $\forall t_1, t_2 \in J_0$,

$$(x(t_2),v(t_2))_{L^2} - (x(t_1),v(t_1))_{L^2}$$
$$+\int_{t_1}^{t_2} \{\mu(x(t),v(t))_{H_0^1} - (x(t),v'(t))_{L^2} + b(x(t),x(t),v(t)) - (f(t)),v(t))_{L^2}\}dt = 0;$$

(3) $\forall t_1, t_2 \in J_0$, *the "energy" equation holds*:

$$\|x(t_2)\|^2_{L^2} - \|x(t_1)\|^2_{L^2} + 2\mu \int_{t_1}^{t_2} \|x(t)\|^2_{H_0^1} dt = 2\int_{t_1}^{t_2} (f(t),x(t))_{L^2} dt.$$

V (*Baiocchi* [1]). *Let* m *be arbitrary and* Ω *"sufficiently smooth." Then every solution* $x(t) \in H^{2/(2+m)}_{\mathrm{loc}}(J_0;\, L^2)$.

VI (*Prouse* [9]). *Let* $m=3$ *and* Ω *of class* C^2. *Then, if* $f(t) \in L^1_{\mathrm{loc}}(J_0;L^2) \cap L^2_{\mathrm{loc}}(J_0;H^{-1/2})$ *and if* $x_0 \in N^\sigma(\frac{1}{2} < \sigma \le 1)$, *there exists* $\bar{t}$, *depending on* x_0 *and* f, *such that the initial value problem*, $x(0)=x_0$, *has, on* $0\vdash\!\dashv\bar{t}$, *one and only one solution*, $x(t)$. *Moreover*, $x(t) \in L^\infty(0\vdash\!\dashv\bar{t};\, N^\sigma) \cap L^2(0\vdash\!\dashv\bar{t};\, H^{1+\sigma})$.

In the above statement, we have denoted by $N^\sigma = [N,N^1]_\sigma (N^0 \equiv N)$ the *intermediate space* between N and N^1, corresponding to the value σ, according to the definition given by Lions [1]. Always according to that definition, it is $H^{1/2} = [L^2,H^1]_{1/2}$; moreover, $H^{-1/2}$ is the dual space of $H^{1/2}$.

Theorem VI establishes, in a neighborhood of $t=0$, a *permanence* property of the solution with respect to the initial data.

(*d*) From now on, we shall consider solutions of (3.6) defined on J; we shall therefore assume that $x(t)$, $f(t)$, $h(t)$ satisfy conditions obtained from (i_1), (i_2), (i_3) by substituting J to J_0^+ or J_0.

Assume that $m=2$. Then the following theorems concerning bounded and a.p. solutions hold (Prouse [4]).

VII. *If* $f(t)$ *is* L^2*-bounded on* J (that is, if

$$\sup_J \|f(t)\|_{L^2} = K < +\infty),$$

then there exists, on J, *at least one* L^2*-bounded solution* $\tilde{x}(t)$: *precisely*,

$$\sup_J \|\tilde{x}(t)\|_{L^2} = M < +\infty,$$

where M *depends only on* K, μ, Ω.

If, moreover, K *is sufficiently small* ($K \le K_0$ *depending only on* μ, Ω), *then the bounded solution* $\tilde{x}(t)$ *is unique.*

VIII. $\sup_J \|f(t)\|_{L^2} \le K_0$, $f(t)$ L^2-S^2 *w.a.p.* $\Rightarrow$ $\tilde{x}(t)$ L^2*-w.a.p. and* L^2-S^2 *a.p.*

IX. $\sup_J \|f(t)\|_{L^2} \le K_1 \le K_0$ (K_1 *sufficiently small, depending only on* μ, Ω), $f(t)$ L^2-S^2 *a.p.* $\Rightarrow$ $\tilde{x}(t)$ L^2*-a.p. and* H_0^1-S^2 *a.p.*

Also in this case the almost-periodicity of the bounded solution $\tilde{x}(t)$ is therefore a consequence of its uniqueness.

(*e*) A study of the Navier-Stokes equation in *more than two dimensions, with a.p. known term*, in which the almost-periodicity of *sufficiently small* solutions is proved, has been made by Fojas [1], who has given the following theorem.

X. *Let $m=3$ and $f(t)$ L^2-a.p. If there exists, on J, a solution $\tilde{x}(t)$ such that, for a certain p, with $3<p\leq+\infty$, we have*

$$\sup_J \|\tilde{x}(t)\|_{L^p} \leq K$$

(K depending only on p and Ω), then $\tilde{x}(t)$ is L^2-a.p.

Finally, we recall that, if $f(t)$ is *periodic*, the proof of the *existence* of at least one periodic solution has been given by Prodi [3], for $m=2$, and, subsequently, by Iudovic [1] and Prouse [3] for any m.

4. A NONLINEAR FUNCTIONAL EQUATION

(*a*) The functional equation we shall consider in this section represents a generalization of the following *nonlinear heat equation* (in operational form):

$$x'(t)-\Delta x(t)+\beta(x(t)) = f(t),$$

where Δ denotes the Laplace operator.

In the sequel, if $\mathscr{H}_1$ and $\mathscr{H}_2$ are two Hilbert spaces, with $\mathscr{H}_1\subset\mathscr{H}_2$ and dense in $\mathscr{H}_2$, we shall denote by $[\mathscr{H}_1,\mathscr{H}_2]_\sigma$ $(0<\sigma<1)$ the Hilbert space *intermediate* between $\mathscr{H}_1$ and $\mathscr{H}_2$, corresponding to the value σ, according to the definition of Lions [1].

Let X_1, X_2, Y be three real Hilbert spaces, with $X_1\subset X_2\subseteq Y$, the embedding of X_1 in X_2 being compact and each space dense in the following one.

Moreover, let $\gamma_1(u,v)$, $\gamma_2(u,v)$ be two bilinear Hermitian forms such that there exist two positive constants c_1, c_2 for which

$$\text{(4.1)}\qquad \begin{aligned} &|\gamma_j(u,v)| \leq c_1\|u\|_{X_j}\|v\|_{X_j}, && \forall u,\, v\in X_j,\\ &\gamma_j(v,v) \geq c_2\|v\|^2_{X_j}, && \forall v\in X_j \qquad (j=1,2). \end{aligned}$$

It is then possible to define in Y two *linear, closed, unbounded operators* Γ_1 and Γ_2, with domains $D(\Gamma_1)$ and $D(\Gamma_2)$, respectively, as follows: an element $u\in X_j$ belongs to $D(\Gamma_j)$ if the linear functional $v\to\gamma_j(u,v)$ is continuous on X_j in the topology induced by Y. *The operator Γ_j is therefore defined by the relation* $\gamma_j(u,v) = (\Gamma_j u,v)_Y$, $\forall v\in X_j$ $(j=1,2)$.

Hence, $D(\Gamma_j)$ is a Hilbert space in the graph norm and, by the assumptions made on $\gamma_j(u,v)$, Γ_j *is linear, self-adjoint,* and *positive.* It is therefore possible to define the powers Γ_j^ν, ν real >0.

Let Λ *be a nonlinear operator from a reflexive Banach space* $K \subseteq Y$ *and dense in* Y *to the dual space* K^*, *with* $\Lambda 0 = 0$; *we assume that there exist* $p > 2$ *and* $\mu > 0$ *such that,* $\forall v \in K$,

$$\|\Lambda v\|_{K^*} \leq \mu(1 + \|v\|_K^{p-1}). \tag{4.2}$$

Moreover, let Γ_3 *be a linear, continuous operator from* $D(\Gamma_1)$ *to* Y.

We shall consider the nonlinear functional equation (Prouse [10])

$$x'(t) + (\Gamma_1 + \Gamma_3)x(t) + \Lambda\Gamma_2 x(t) = f(t). \tag{4.3}$$

For the sake of simplicity, we shall not give here in detail all the assumptions that have to be made in order that the theorems we shall state hold. The main hypotheses can, however, be roughly summarized by saying that we shall assume that Γ_1 and Γ_2 are *permutable* and that Γ_3 is "*weaker*" than Γ_1; moreover, Λ is weakly continuous from finite-dimensional subspaces of K to K^* and satisfies the relations

$$\langle \Lambda v, v \rangle \geq c_1 \|v\|_K^2 - c_2,$$

$$\langle \Lambda v_1 - \Lambda v_2, v_1 - v_2 \rangle \geq -c_3 \|v_1 - v_2\|_Y^2,$$

where c_1, c_2, c_3 are positive constants; $\langle\ \rangle$ denotes the duality between K and K^*.

Let us observe that Lions and Strauss [1] have considered the initial value problem

$$x(0) = x_0 \tag{4.4}$$

for an equation of the type:

$$x'(t) + \Gamma(t)x(t) + \Lambda(t)x(t) = f(t),$$

which, formally, contains (4.3). The existence and uniqueness theorem for the solution of the initial value problem that we shall state does not, however, follow from that proved by those authors.

Setting $V = D((\Gamma_1, \Gamma_2)^{1/2})$, $Z = [D(\Gamma_2), D(\Gamma_1)^*]_{1/2}$, $J_0 = 0 \vdash\!\!\!- +\infty$, and assuming that $f(t) \in L^r_{\text{loc}}(J_0; K^*)$, we shall say that $x(t)$ is a *solution* on J_0 of (4.3) if:

(1) $x(t) \in L^\infty_{\text{loc}}(J_0; X_2) \cap L^2_{\text{loc}}(J_0; V)$, $\Gamma_2 x(t) \in L^p_{\text{loc}}(J_0; K)$;

(2) $x(t)$ *satisfies* (4.3) *a.e. on* J_0, *in the sense of distributions with values in* $Z + K^*$.

Observe that, if we set $X_2 = Y = K = L^2$, $X_1 = H_0^1$, $\Gamma_1 + \Gamma_3 = -\Delta$, $\Gamma_2 = I$, $\Lambda = 0$ (hence $V = D((\Gamma_1\Gamma_2)^{1/2}) = D((-\Delta)^{1/2}) = H_0^1$), then (4.3) reduces to the heat equation and the definition of solution indicated above coincides with that usually given (in the weak sense) for the heat equation (see, for example, Lions [2]).

(*b*) We now state the main theorems which hold for the solutions of (4.3).

I. *If* $f(t) \in L^r_{loc}(J_0;K^*)$, *there exists on* J_0 *one, and only one, solution of* (4.3), $x(t)$, *satisfying the initial condition* (4.4), *where* x_0 *is an arbitrary element of* X_2.

II. *If*

$$\sup_J \left\{\int_0^1 \|f(t+\eta)\|^r_{K^*} d\eta\right\}^{1/r} < +\infty,$$

then there exists one, and only one, bounded solution, $\tilde{x}(t)$, *on* J, *such that*

$$\sup_J \|\tilde{x}(t)\|_{X_2} < +\infty, \qquad \sup_J \left\{\int_0^1 \|\tilde{x}(t+\eta)\|^2_V d\eta\right\}^{1/2} < +\infty,$$

$$\sup_J \left\{\int_0^1 \|\Gamma_2 \tilde{x}(t+\eta)\|^p_K d\eta\right\}^{1/p} < +\infty.$$

Moreover, if $f(t)$ *is periodic with period* Θ, $\tilde{x}(t)$ *is also periodic, with the same period.*

III. $f(t)$ K^*-S^r *w.a.p.* $\Rightarrow$ $\tilde{x}(t)$ X_2-S^2 *a.p.*

IV. $f(t)$ K^*-S^r *a.p.* $\Rightarrow$ $\tilde{x}(t)$ X_2-*a.p. and* V-S^2 *a.p.*

(*c*) We shall now give an example of an equation for which the theorems stated above hold. We shall also, in this case, state the assumptions that were not given in detail for the general equation (4.3). Let $Y = L^2(\Omega) = L^2$ (Ω open, connected, and bounded set of R^m) and consider the operators

$$\Gamma_1 = -\Delta, \qquad \Gamma_3(\zeta) = \sum_1^m {}_j\, a_j(\zeta) \frac{\partial}{\partial \zeta_j} + a_0(\zeta),$$

$$\Gamma_4 = \sum_{j,k}^{1\ldots m} b_{jk} \frac{\partial^2}{\partial \zeta_j \partial \zeta_k} + b_0,$$

with $b_{jk} = b_{kj} = \text{const}$, $b_0 = \text{const}$, $a_j(\zeta) \in L^\infty(\Omega)(j = 0, \cdots, m)$. We assume, moreover, that, $\forall v \in H^1_0(\Omega) = H^1_0$,

$$\langle \Gamma_4 v, v\rangle_{H^1_0} \geq \mu \|v\|^2_{H^1_0}, \qquad (\mu > 0),$$

where the symbol $\langle\ \rangle_{H^1_0}$ denotes the duality between the space H^1_0 and $H^{-1} = \Gamma_1 H^1_0 = \Gamma_4 H^1_0$.

Let $\beta(\eta)$ be a real continuous function, satisfying locally a Lipschitz condition, defined on J and such that $\beta(0) = 0$; assume that there exist a number $p \geq 2$ and three positive constants c_1, c_2, c_3 such that

$$c_1|\eta|^p \leq \eta\beta(\eta) \leq c_2|\eta|^p \quad \text{when } |\eta| \geq \bar{\eta} \text{ sufficiently large},$$

$$\frac{\beta(\eta_1) - \beta(\eta_2)}{\eta_1 - \eta_2} \geq -c_3, \qquad \forall \text{ pair } (\eta_1, \eta_2) \in J.$$

We consider now, for $0 \leq \sigma < 1$ and $t \in J_0$, the equation

$$\frac{\partial x(t,\zeta)}{\partial t} + (\Gamma_1 + \Gamma_3(\zeta))x(t,\zeta) + \beta(\Gamma_4^\sigma x(t,\zeta)) = f(t,\zeta)$$

with the *initial* and *boundary conditions*

$$x(0,\zeta) = x_0(\zeta) \in [L^2, H_0^1]_\sigma, \qquad x(t,\zeta)|_{\zeta \in \partial\Omega} = 0.$$

This equation is a special case of (4.3), setting $X_1 = H_0^1$, $X_2 = [L^2, H_0^1]_\sigma$, $K = L^p(\Omega) = L^p$ (and, consequently, $V = [L^2, H^2 \cap H_0^1)]_{(1+\sigma)/2}$, $K^* = L^r$).

Applying the theory stated above, it can then be shown that, if the coefficients $a_j(\zeta)$ are "sufficiently small" and if $\partial\Omega$ is of class C^{2s}, with $s = [m(p-2)/4p] + 1$, the existence and uniqueness theorem for the initial value problem and the existence theorem of a bounded (and, eventually, periodic) solution hold.

If, moreover, the constant c_3 is "sufficiently small" (in particular: *if $\beta(\eta)$ is nondecreasing*), then *the bounded* (or *periodic*) *solution is unique and, if $f(t)$ is a.p., the almost-periodicity theorems III and IV hold.*

BIBLIOGRAPHY

Amerio, L.

[1] Soluzioni quasi-periodiche, o limitate, di sistemi differenziali quasi-periodici, o limitati, *Ann. di Mat.* **39**, 1955.

[2] Problema misto e quasi-periodicità per l'equazione delle onde non omogenea, *Ann. di Mat.* **29**, 1960.

[3] Quasi-periodicità degli integrali ed energia limitata dell' equazione delle onde con termine noto quasi-periodico, I, II, III, *Rend. Acc. Naz. Lincei* **28**, 1960.

[4] Sull'integrazione delle funzioni quasi-periodiche a valori in uno spazio hilbertiano, *Rend. Acc. Naz. Lincei* **38**, 1960.

[5] Funzioni debolmente quasi-periodiche, *Rend. Sem. Mat. Padova* **30**, 1960.

[6] Sull'equazione delle onde con termine noto quasi-periodico, *Rend. di Mat. Univ. di Roma* **19**, 1960.

[7] Sulle equazioni differenziali quasi-periodiche astratte, *Ric. di Mat.* **9**, 1960; Ancora sulle equazioni differenziali quasi-periodiche astratte, *Ric. di Mat.* **10**, 1961.

[8] Sulle equazioni lineari quasi-periodiche negli spazi hilbertiani I, II, *Rend. Acc. Naz. Lincei* **31**, 1961.

[9] Sull'integrazione delle funzioni quasi-periodiche astratte, *Ann. di Mat.* **53**, 1961.

[10] Soluzioni quasi-periodiche delle equazioni lineari iperboliche quasi-periodiche, *Rend. Acc. Naz. Lincei* **33**, 1962.

[11] Sull'integrazione delle funzioni $l^p\{X_n\}$-quasi-periodiche, con $1 \leq p < +\infty$, *Ric. di Mat.* **12**, 1963.

[12] Soluzioni quasi-periodiche di equazioni quasi-periodiche negli spazi hilbertiani, *Ann. di Mat.* **61**, 1963.

[13] Su un teorema di minimax per le equazioni differenziali astratte, *Rend. Acc. Naz. Lincei* **34**, 1963.

[14] Sul teorema di approssimazione delle funzioni quasi-periodiche, *Rend. Acc. Naz. Lincei* **34**, 1963; Ancora su un teorema di approssimazione delle funzioni quasi-periodiche, *Rend. Acc. Naz. Lincei* **36**, 1964.

[15] Solutions presque-périodiques d'équations fonctionnelles dans les espaces de Hilbert, *Coll. sur l'Analyse fonctionelle*, Liège, 1964.

[16] Abstract almost-periodic functions and functional equations, *Boll. U.M.I.* **20**, 1965.

[17] Almost-periodic solutions of the equation of Schrödinger type, I, II, *Rend. Acc. Naz. Lincei* **43**, 1967.

Amerio, L. and Prouse, G.

[1] On the nonlinear wave equation with dissipative term discontinuous with respect to the velocity, I, II, *Rend. Acc. Naz. Lincei* **44**, 1968.

[2] Bounded or almost-periodic solutions of a nonlinear wave equation, I, II, *Rend. Acc. Naz. Lincei* **44**, 1968.

Baiocchi, C.

[1] Sulle soluzioni del sistema di Navier-Stokes in dimensione n, I, II, III, *Rend. Acc. Naz. Lincei* **40–41**, 1966.

Besicovic, A. S.

[1] *Almost-Periodic Functions*, Dover Publications, New York, 1958.

BOCHNER, S.

[1] Abstrakte Fastperiodische Funktionen, *Acta Math.* **61**, 1933.

[2] Fastperiodische Lösungen der Wellengleichung, *Acta Math.* **62**, 1934.

[3] Almost-periodic solutions of the inhomogeneous wave equation, *Proc. Nat. Acad. Sci.* **46**, 1960.

[4] Uniform convergence of monotonic sequences of functions, *Proc. Nat. Acad. Sci.* **47**, 1961.

BOCHNER, S. AND VON NEUMANN, J.

[1] Almost-periodic functions of groups, II, *Trans. Am. Math. Soc.* **37**, 1935.

[2] On compact solutions of operational differential equations, *Ann. Math.* **36**, 1935.

BOHR, H.

[1] Zur Theorie der fastperiodische Funktionen, *Acta Math.* **46**, 1925.

[2] *Fastperiodische Funktionen*, Springer, Berlin, 1932.

CINQUINI, S.

[1] *Funzioni Quasi-Periodiche*, Quaderni matematici della Sc. Norm. Sup. di Pisa, **19**, 1948–1949.

CORDUNEANU, C.

[1] *Functii Aproape—Periodice*, Editura Academiei Republicii Populare Romîne, 1961.

[2] *Almost Periodic Functions*, Interscience, New York, 1968.

DAY, M.

[1] *Normed Linear Spaces*, Springer, Berlin, 1958.

DOLCHER, M.

[1] Su un criterio di convergenza uniforme per le successioni monotone di funzioni quasi-periodiche, *Rend. Sem. Mat. Padova* **34**, 1964.

DUNFORD, N. AND SCHWARTZ, J.

[1] *Linear Operators*, Interscience, New York, 1958.

EBERLEIN, W. E.

[1] Abstract ergodic theorems and weak almost-periodic functions, *Trans. Am. Math. Soc.* **67**, 1949.

[2] The spectrum of weakly almost-periodic functions, *Mich. J. Math.* **3**, 1956.

FAVARD, J.

[1] *Leçons sur les fonctions presque-périodiques*, Gauthier-Villars, Paris, 1933.

FOJAS, C.

[1] Essais dans l'étude des solutions des équations de Navier-Stokes dans l'espace. L'unicité et la presque-périodicité des solutions petites, *Rend. Sem. Mat. Padova* **32**, 1962.

FOJAS, C. AND ZAIDMAN, S.

[1] Almost-periodic solutions of parabolic systems, *Ann. Scuola Norm. Sup. Pisa* **15**, 1962.

GÜNZLER, H.

[1] Fastperiodische Lösungen linearer hyperbolischer Differentialgleichungen, *Math. Zeitschr.* **71**, 1959.

[2] Hyperbolic differential equations, *Rend. Sem. Mat. Fis. Milano* **34**, 1964.

[3] Vector valued functions on semigroups with almost-periodic differences, *Rend. Acc. Naz. Lincei* **42**, 1967.

[4] Beschränktheitseigenschaften für Lösungen nichtlinearer Wellengleichungen, *Math. Ann.* **167**, 1966.

[5] Integration of almost-periodic functions, *Math. Z.* **102**, 1967.

[6] Means over countable semigroups and almost-periodicity in l^1, *J. Reine Angew. Math.* **232**, 1968.

GÜNZLER, H. AND ZAIDMAN, S.

[1] Almost-periodic solutions of abstract differential equations, *Math. Inst. Univ. Göttingen, Sem. W. Maak, Bericht No.* **24**, 1968.

[2] Almost-periodic solutions of certain abstract differential equations, *Journ. Math. Mech.* **19**, 1969.

HILLE, E. AND PHILLIPS, R. S.

[1] *Functional analysis and semi-groups*, A.M.S. Colloquium Publications **31**, 1957.

HOPF, E.

[1] Über di Anfangswertaufgabe für die hydrodinamischen Grundgleichungen, *Math. Nach.* **4**, 1951.

IUDOVIC, V.

[1] Periodic solutions of a viscous incompressible fluid (in Russian), *Dokl. Akad. Nauk SSSR* **130**, 1960.

KIESELEV, A. AND LADYŽENSKAJA, O.

[1] On the existence and uniqueness of the solution of the nonstationary problem for an incompressible viscous fluid (in Russian), *Isv. Akad. Nauk* **21**, 1957.

KOPEC, J.

[1] On linear differential equations in Banach spaces, *Zes. Nauk. Univ. Mickiewicza Mat. Chem.* **1**, 1957.

[2] On vector-valued almost-periodic functions, *Ann. Soc. Pol. Math.* **79**, 1962.

LADYŽENSKAJA, O.

[1] *The Mixed Problem for Hyperbolic Equations* (in Russian), Moscow, Leningrad, 1953.

[2] Solutions in the large of the boundary value problem for the Navier-Stokes equation in two space variables, *Comm. Pure Appl. Math.* **12**, 1959.

[3] *The Mathematical Theory of Viscous Incompressible Flow*, Gordon and Breach, New York, 1964.

LERAY, J.

[1] Etude des diverses équations intégrals non linéaires et de quelques problèmes qui pose l'hydrodinamique, *J. Math. Pures et Appl.* **12**, 1933.

LEVITAN, B. M.

[1] *Almost-Periodic Functions* (in Russian), Moscow, 1953.

[2] Integration of almost-periodic functions with values in Banach spaces (in Russian), *Izv. Akad. Nauk SSSR* **30**, 1966.

LIKOV, V.

[1] On the existence of almost-periodic solutions of operational differential equations (in Russian), *Collection of scientific papers, Polytechnic of Vladimir*, 1969.

LIONS, J. L.

[1] Espaces intérmediaires entre espaces hilbertiens et applications, *Bull. Soc. Math. Phys. Roumainie* **50**, 1958.

[2] *Equations Différentielles Opérationnelles et Problèmes aux Limites*, Springer, Berlin, 1962.

LIONS, J. L. AND MAGENES, E.

[1] *Problèmes aux Limites non Homogènes et Applications*, Dunod, Paris, 1968.

LIONS, J. L. AND PRODI, G.

[1] Un théorème d'existence et d'unicité dans les équations de Navier-Stokes en dimension 2, *C.R. Acad. Sci.* **248**, 1959.

LIONS, J. L. AND STRAUSS, W. A.

[1] Some non-linear evolution equations, *Bull. Soc. Math. France* **93**, 1965.

MAAK, W.

[1] *Fastperiodische Funktionen*, Springer, Berlin, 1950.

MUCKENHOUPT, C. F.
[1] Almost-periodic functions and vibrating systems, *J. Math. Phys.* **8**, 1929.

NAGY, B.
[1] Vibrations d'une corde non homogène, *Bull. Soc. Math. France* **75**, 1947.

PRODI, G.
[1] Soluzioni periodiche di equazioni alle derivate parziali di tipo parabolico non-lineari, *Riv. Mat. Univ. Parma* **5**, 1952.
[2] Qualche risultato riguardo alle equazioni di Navier-Stokes nel caso bidimensionale, *Rend. Sem. Mat. Padova* **30**, 1960.
[3] Rassegna di ricerche intorno alle equazioni di Navier-Stokes. *Quaderno n.* 2, *Università di Trieste.*
[4] Soluzioni periodiche dell'equazione delle onde con termine dissipativo non lineare, *Rend. Sem. Mat. Padova* **36**, 1966.

PROUSE, G.
[1] Analisi di alcuni classici problemi di propagazione, *Rend. Sem. Mat. Padova* **32**, 1962.
[2] Sulle equazioni differenziali astratte quasi-periodiche secondo Stepanov, *Ric. di Mat.* **9**, 1962.
[3] Soluzioni periodiche dell'equazione di Navier-Stokes, *Rend. Acc. Naz. Lincei* **35**, 1963.
[4] Soluzioni quasi-periodiche dell'equazione di Navier-Stokes in due dimensioni, *Rend. Sem. Mat. Padova* **33**, 1963.
[5] Su un esempio concernente l'equazione delle onde, *Rend. Acc. Naz. Lincei* **37**, 1964.
[6] Soluzioni periodiche dell'equazione delle onde non omogenea con termine dissipativo quadratico, *Ric. di Mat.* **13**, 1964.
[7] Soluzioni quasi-periodiche dell'equazione non omogenea della membrana vibrante con termine dissipativo quadratico, I, II, *Rend. Acc. Naz. Lincei* **37**, 1964.
[8] Soluzioni quasi-periodiche dell'equazione non omogenea delle onde con termine dissipativo non lineare, I, II, III, IV, *Rend. Acc. Naz. Lincei* **38–39**, 1965.
[9] Un teorema di esistenza, unicità e regolarità per il sistema di Navier-Stokes nello spazio, *Rend. Acc. Naz. Lincei* **42**, 1967.
[10] Periodic or almost-periodic solutions of a non-linear functional equation, I, II, III, IV, *Rend. Acc. Naz. Lincei* **43–44**, 1967–68.

RICCI, M. L.
[1] Soluzioni quasi-periodiche di un'equazione a derivate parziali del tipo di Fuchs, *Ann. di Mat.* **79**, 1966.

RICCI, M. L. AND RIZZONELLI, P.
[1] Sulle funzioni l^1-quasi-periodiche, *Rend. Ist. Lombardo* **95**, 1961.

RICCI, M. L. AND VAGHI, C.
[1] Soluzioni quasi-periodiche di un'equazione funzionale, *Ric. di Mat.* **13**, 1964.

RIESZ, F. AND NAGY, B.
[1] *Functional Analysis*, Blackie, London, 1956.

SOBOLEV, S. L.
[1] On a theorem in functional analysis (in Russian), *Mat. Sbornik* **46**, 1938.
[2] Sur la presque-périodicité des solutions de l'équation des ondes, *Dokl. Akad. Nauk SSSR* **48**, 1945.

STRAUSS, A. W.
[1] On continuity of functions with values in certain Banach spaces, *Pacific J. Math.* **19**, 1966.

TORELLI, G.

[1] Un complemento ad un teorema di J. L. Lions sulle equazioni differenziali astratte del secondo ordine, *Rend. Sem. Mat. Padova* **34**, 1964.

VAGHI, C.

[1] Soluzioni C^0-quasi-periodiche dell'equazione non omogenea delle onde, *Ric. di Mat.* **12**, 1963.

[2] Su un'equazione iperbolica con coefficienti periodici e termine noto quasi-periodico, *Rend. Ist. Lombardo* **100**, 1966.

[3] Sul comportamento asintotico delle soluzioni di equazioni non lineari di tipo parabolico, I, II, *Rend. Acc. Naz. Lincei* **46**, 1966.

[4] Soluzioni limitate, o quasi-periodiche, di un'equazione di tipo parabolico non lineare, *Boll. U.M.I.* **4–5**, 1968.

[5] Su un' equazioni parabolica con termine quadratico nella derivata z_x, *Rend. Ist. Lombardo* **103**, 1969.

VASCONI, A.

[1] Sull'integrazione delle funzioni quasi-periodiche secondo Stepanov, negli spazi di Clarkson, *Rend. Ist. Lombardo* **95**, 1961.

[2] Sull'equazione delle onde con termine noto quasi-periodico secondo Stepanov, *Rend. Ist. Lombardo* **96**, 1962.

YOSHIDA, K.

[1] *Functional Analysis*, Springer, 1966.

ZAIDMAN, S.

[1] Sur la presque-périodicité des solutions de l'équation non homogène des ondes, *J. Math. Mech.* **8**, 1959.

[2] Solutions presque périodiques dans le problème de Cauchy pour l'équation non homogène des ondes, I, II, *Rend. Acc. Naz. Lincei* **30**, 1961.

[3] Solutions presque-périodiques des équations hyperboliques, *Ann. Ec. Norm. Sup.* **79**, 1962.

[4] Soluzioni limitate o quasi-periodiche dell'equazione del calore non omogenea, I, II, *Rend. Acc. Naz. Lincei* **32**, 1962.

[5] Quasi-periodicità per un'equazione operazionale del primo ordine, *Rend. Acc. Naz. Lincei* **35**, 1963.

[6] Quasi-periodicità per l'equazione di Poisson, *Rend. Acc. Naz. Lincei* **34**, 1963.

[7] Soluzioni quasi-periodiche per alcune equazioni differenziali in spazi hilbertiani, *Ric. di Mat.* **3**, 1964.

[8] Soluzioni limitate, o quasi-periodiche, dell'equazione di Poisson, *Ann. di Mat.* **54**, 1964.

ZIKOV, V.

[1] Abstract equations with almost-periodic coefficients (in Russian), *Dokl. Akad. Nauk SSSR* **166**, 1965.

[2] Almost-periodic solutions of differential equations in Hilbert space (in Russian), *Dokl. Akad. Nauk SSSR* **166**, 1965.

INDEX